CAMBRIDGE LIBRARY COLLECTION

Books of enduring scholarly value

Physical Sciences

From ancient times, humans have tried to understand the workings of
the world around them. The roots of modern physical science go back to
the very earliest mechanical devices such as levers and rollers, the mixing
of paints and dyes, and the importance of the heavenly bodies in early
religious observance and navigation. The physical sciences as we know them
today began to emerge as independent academic subjects during the early
modern period, in the work of Newton and other 'natural philosophers',
and numerous sub-disciplines developed during the centuries that followed.
This part of the Cambridge Library Collection is devoted to landmark
publications in this area which will be of interest to historians of science
concerned with individual scientists, particular discoveries, and advances in
scientific method, or with the establishment and development of scientific
institutions around the world.

The Life of Roger Langdon

First published in 1909, this autobiography details the astonishing life
of Roger Langdon (1825–94), a country station-master and amateur
astronomer. Langdon's life is a remarkable story of self-education and
determination: he started work as a farmer's boy at the age of eight, ran away
from the home to work for a shipowner in Jersey at fourteen, and was then
employed by a blacksmith, canvas manufacturers, and a solicitor before
finding work with the Great Western Railway. Langdon was from an early
age interested in astronomy, and eventually constructed four telescopes
and his own observatory. He developed his own method for photographing
the moon and the transit of Venus, and presented a paper to the Royal
Astronomical Society, which is included in the appendices. Langdon died
before completing his autobiography, and the latter chapters on his scientific
achievements and final years were completed by his daughter Ellen.

Cambridge University Press has long been a pioneer in the reissuing of out-of-print titles from its own backlist, producing digital reprints of books that are still sought after by scholars and students but could not be reprinted economically using traditional technology. The Cambridge Library Collection extends this activity to a wider range of books which are still of importance to researchers and professionals, either for the source material they contain, or as landmarks in the history of their academic discipline.

Drawing from the world-renowned collections in the Cambridge University Library, and guided by the advice of experts in each subject area, Cambridge University Press is using state-of-the-art scanning machines in its own Printing House to capture the content of each book selected for inclusion. The files are processed to give a consistently clear, crisp image, and the books finished to the high quality standard for which the Press is recognised around the world. The latest print-on-demand technology ensures that the books will remain available indefinitely, and that orders for single or multiple copies can quickly be supplied.

The Cambridge Library Collection will bring back to life books of enduring scholarly value (including out-of-copyright works originally issued by other publishers) across a wide range of disciplines in the humanities and social sciences and in science and technology.

The Life of Roger Langdon

Told by Himself,
with Additions by his Daughter

ROGER LANGDON
ELLEN LANGDON

CAMBRIDGE UNIVERSITY PRESS

Cambridge, New York, Melbourne, Madrid, Cape Town, Singapore,
São Paolo, Delhi, Dubai, Tokyo, Mexico City

Published in the United States of America by Cambridge University Press, New York

www.cambridge.org
Information on this title: www.cambridge.org/9781108021647

© in this compilation Cambridge University Press 2010

This edition first published 1909
This digitally printed version 2010

ISBN 978-1-108-02164-7 Paperback

THE LIFE OF ROGER LANGDON

THE LIFE OF
ROGER LANGDON

TOLD BY HIMSELF

With Additions by his Daughter
Ellen

LONDON
ELLIOT STOCK
62, PATERNOSTER ROW, E.C

"PROGRESS OF ASTRONOMY"

[*From "Whitaker's Almanack" for* 1895, *under the heading "Progress of Astronomy."*]

Mr. Langdon, station-master at Silverton, on the Great Western Railway, a self-taught astronomer, died on July 18, 1894. Mr. Langdon made in his spare hours an 8-inch silver-on-glass mirror, grinding it on a machine of his own construction. In 1872 he contributed a paper to the *Monthly Notices* of the Royal Astronomical Society on " The Markings of Venus."

PREFACE

THE writing of this foreword to the biography of the late Mr. Roger Langdon should have devolved upon one of the notable personages who had an admiration for him and his work, but unhappily they have all, or nearly all, passed away. Unquestionably the person best fitted for the task would have been the late Rev. H. Fox Strangways, rector of Silverton during the period when Mr. Langdon acted as station-master there. They had a very cordial liking and respect for each other, and Mr. Strangways could doubtless have imparted a personal and intimate touch to this preface which would have been very valuable.

When Miss Ellen Langdon desired me to undertake this portion of the work I felt honoured, though diffident. A feeling that it was my clear duty to pay any mark of respect I could to the memory of this worthy man decided me to accept her invitation.

My acquaintance with Mr. Langdon dates back to a few years before his death when my father was general manager of the Great Western Railway and Mr. Langdon was still at work at Silverton. My father's attention had been called to the personality and attainments of the

Silverton station-master, and as I was at that time
doing a little journalism in odd moments it was suggested
that I should run down and write something for the
Great Western Magazine, which I was very pleased to do.
At that little wayside station just on the London side
of Exeter I therefore found myself one summer afternoon. The village of Silverton, distant two miles from
the station, was not visible, and the principal features
in the immediate vicinity were the station-master's house,
with the front garden between it and the station, and in
the front garden a circular iron building with a cone-
shaped revolving roof, which, I found, was an observatory
sheltering a telescope for celestial observation.

The tall, slightly stooping, white-bearded old station-
master at once arrested attention. A dignified, patriar-
chal type of man, with a kindly, pleasant and simple
manner, he was evidently much averse to all forms of
affectation and cant. I was quickly made welcome and
introduced to his wife and well-ordered home.

We were immediately on excellent terms. I remember
the eager pride with which he showed me his beloved
telescope and its mounting and accessories, including the
sidereal clock, and how I gazed under his direction at the
heavenly objects which the night disclosed. The evening we spent together was a very memorable one. Mr.
Langdon recounted the hardships and adventures of his
career, and gave me an insight into the manifold difficulties
and obstacles he had overcome in attaining the means
of observing the celestial bodies in which he took so
absorbing an interest. He also displayed for my amuse-

ment the ingenious church with chimes and other works of his hands.

It is distinctly to be regretted that his autobiography ceases before the period when he made his four telescopes. His own account of his trials and difficulties and of the indefatigable inventive genius he showed in grappling with them would have been most instructive. His achievements become very impressive when his environment and paucity of means are remembered.

Long hours of duty at a little country station, the support and clothing of himself, his wife, and eight children who required to be educated and placed out in the world—all accomplished on a weekly wage, which from his marriage to old age averaged only 30s., and was in the earlier years much less—would have been enough to exhaust the energy and resources of any ordinary man. Nevertheless Mr. Langdon found time and means to learn French, Greek, and Shorthand, to amuse his family and neighbours with lantern lectures, and to make and use effectively four telescopes, so that eventually his reputation spread to the Royal Astronomical Society, before which he read a paper on his discoveries and observations. Bear in mind that money was so scarce that he was practically reduced to make everything, even his tools, with his own hands from the crude materials, groping his way through the mists of uncertainty and disappointment to the haven of ultimate success.

He was fortunate in his marriage, or he would probably never have succeeded as he did. He always referred to his wife as an inestimable blessing, and was, by her help, as free from home cares as a man with so small an income and eight children could be. The widow of the late rector of Silverton bears testimony to the virtues and many good works of this estimable couple. Their children rise up and call them blessed. Their character and example even in this small locality and limited sphere must have been of very marked value.

The career of Roger Langdon provides for all of us a striking illustration of what force of character will accomplish even in the humblest surroundings and in the face of the most serious obstacles. Such men working persistently onwards and upwards with such slight recognition and encouragement are the real heroes of life, and their memory should be kept green for the benefit of those who come after them.

H. CLIFTON LAMBERT.

CONTENTS

CHAPTER I

"WHY WAS I BORN?"

As earth's pageant passes by,
Let reflection turn thine eye
Inward, and observe thy breast;
There alone dwells solid rest.
That's a close immured tower
Which can mock all hostile power;
To thyself a tenant be
And inhabit safe and free.
Say not that the house is small
Girt up in a narrow wall
The infinite Creator can
Dwell there—and may not man?
There content make thine abode
With thyself and with thy God.

I HAVE no distinct recollection of my birth, although I believe I was a prominent actor in the performance. The very first thing, or rather, circumstance that I remember, was the birth of my sister, when I was two years and five months old. Old Nanny Holland, who did duty as midwife, nurse and housekeeper, used to allow me to go out and play with the water and dabble in the mud; then she would call me in and smack me well and call me bad names, and shut me under the stairs until my pinafore was dry. I can quite well remember crying and asking myself, " Why was I born ? " especially as old Nanny paid greater attention to me in this respect, than to any of my older brothers. Then, as I grew older, there was my father who thoroughly believed that the stick was a cure for all complaints, and acting upon King Solomon's advice, never spared the rod. On these occasions, I always asked myself the question, " Why was I born ? "

As soon as old Nanny had gone out of the house, I asked my mother if it was likely that old Nanny would bring another baby next week ; and when my kind and loving mother stroked my hair, and smiled and said " No," I was very soon out in the lane making bricks and building houses with mud. My mother did not smack me for this as old Nanny had done, but she would call me and speak to me about making myself dirty, and somehow, whenever she spoke she was always obeyed. She used to have me by her knee and teach me Dr. Watts's hymns. I have lived to hear those hymns scoffed at, but I still think they might do good to some young people. Now at the age of fifty I take great delight in the study of science and astronomy. Who shall say that my dear good mother did not lay the foundation stone, and set my young mind thinking of the wonderful works of God, by teaching me—

> I sing the almighty power of God
>   That made the mountains rise,
> That spread the flowing seas abroad
>   And built the lofty skies.
>
> I sing the wisdom that ordained
>   The sun to rule the day.

The moon shines full at His command
And all the stars obey.

This hymn, and other precepts taught by my gentle mother, sank deep into my mind, and set me thinking and pondering over the works of God, and led me to ask all sorts of questions, and I might say that I received all sorts of answers, which made me still more inquisitive, until my father would tell me to hold my tongue. I do not wish it to be understood that my father was a wrong-headed man, far from it ; for I am sure that he possessed some of the finest qualities that adorn human nature. He possessed, in the very highest degree, the qualities of truth, justice, honour, and honesty of purpose ; he considered it an exceedingly bad practice to owe anything to anybody, so he rose very early in the morning and took rest late that he might maintain his children, in what he termed "poor independence." Moreover, he being the parish clerk and Sunday school master—there was no week-day school—he had a very high veneration for the Church. He was also choir master and organist. Therefore he was a power in the village, and used his stick accordingly. Woe to any bellringer who thoughtlessly entered the door of the church, without removing his hat from his head. "How dare you," he would say, "enter the sanctuary of the Lord in that heathenish manner ? " and the men I know very highly respected him, and obeyed his orders without a murmur. He would never allow cider, which was the drink of the country, to be brought inside the church gate ; it was consecrated ground and was not to be defiled

He was like Job in one thing, he was the father of seven sons and three daughters.

The state of England at that time was very bad indeed, and the poor were really oppressed, especially in our remote part of the country. Well, my father had enough to do to make both ends meet, and how he and my mother slaved and toiled to keep out of debt! My brothers and myself were sent to work at a very early age, at whatever we could get, and at this period, when the oppression was so great, I was always asking myself, " Why was I born ? "

In the year 1829, when I was four years of age, my father and mother had not heard of Dr. Jenner, and his plan of vaccination. If they had they would have surely fallen in with the idea, and would have acted upon it. It was the custom in those days that whenever small-pox made its appearance in the village, the mother of a family would take one of her children to the infected house, and place her healthy child in the bed of the person who had the malady. This was done so that the infection should not come upon her family unawares, but that she might be somewhat in a position to receive it, and with a little judicious management, generally to keep the disease under subjection ; that is to say, she could generally manage so that only one of her children should be down with the small-pox at one time. Whereas, if she allowed the infection to come upon her in its natural course, probably all her children would be down at once with the disease.

Now there was a boy who was said to be dying of

small-pox, and whether it was ignorance, or superstition, or a combination of both, I do not know, but it
was considered best, to let your children catch the
small-pox from those who were suffering most violently.
Accordingly I was taken to the house where the boy lay
dying, and there I was partly undressed and placed in
the cradle by the side of the boy, and I was to stay there
until I got warm and comfortable. As far as my own
thoughts went in the matter I thought it very good fun,
especially as when I was ill I should be out of the way of
the stick at any rate. But while I was thinking over
these matters, who should stalk into the room but old
Nanny Holland. Nanny was a sort of oracle in the
village, besides being a kind of quack doctor, and what
with her superior cunning, and evil temper, always
excited more or less with gin, she held most of the poor
women under her thumb, and when she approached the
cradle where we were lying, I thought she looked more
evil than usual. She looked at the cradle, then at the
boy's mother, and said, " Why don't you let the cheil
(*child*) die ? He can't die shut up in an infernal crile
like this." And thereupon she dragged me out, and
put me down, by no means lightly, upon the floor ; she
then tore away the foot of the cradle, so that the boy's
feet could extend further down, and he was a corpse
directly.

It appears from Nanny's theory, that although the
child was in the agony of death, and with the last pang
upon him, yet the vital spark could not part from him,
until his crib was lengthened sufficiently to allow his feet

to stretch downward without hindrance. I have sometimes thought that perhaps old Nanny was more than half right in her theory.

Now, I cannot tell whether the virus of the boy's small-pox was too far spent, or whether I was an extraordinarily healthy subject, or whether perhaps old Nanny frightened me, but certain it is, I did not catch the smallpox. Therefore there was but one alternative, and that was, that I must be inoculated, or, as the villagers expressed, it "knockle-headed." As soon as I discovered this I really began to quake with fear, and to wonder why I was born. Not that I feared the operation itself, as I had seen it performed on others, but I dreadfully feared the doctor who would perform upon me. I had not long to wait before my suspicions and fears were brought to a climax, for my mother took me off to Nanny Holland.

Nanny soon began to see about "knockle-heading" the children, and when she turned to me first, and I saw her coming towards me, with her surgical knife, my hair stood on end with fright. Where she obtained the virus from I do not know, but she clawed hold of my arm, and stabbed a stocking needle through the skin, and lifting the skin upwards at the same time with a razor in her hand, cut a piece, about the size of a threepenny bit, three parts off, a bit of the skin being left in the way of a hinge ; then with the point of an old knife, she plastered some matter into the wound, just as you might see a painter stopping a hole in a board with putty ; then she replaced the slice of skin with the following caution

"Now, youngster, if you scratch that off, I'll kill thee."
My little sister was put through the same process, and
Louisa Gard, a little chubby happy cherub of about four
years of age, and a constant playmate of ours, was also
operated upon.

In due course old Nanny's "matter" began to work.
My sister was very ill with small-pox, and so also was
little Louisa. As for myself, I had it very slightly, in
fact no one but my mother knew that I had the malady
upon me. My sister got well in time, but of course the
small-pox left its marks severely upon her. Poor little
Louisa never rallied ; or if she got over the small-pox,
she had croup, which was too much for her, and she crossed
over into the Land of Beulah.

Louisa and my sister and myself had attended the
Sunday school, for there was no week-day school. I asked
mother if Louisa would come back, and she said "No, but
if you are a good boy, you will someday go where she
is gone." Then I would go out and look up at the stars,
and wonder if I should see Louisa flitting about from
star to star, but my mother said, "No, you will not see
her there, but you will meet her again at the last day;
and if you grow up to be a good man, you will hear the
Great Judge say, 'Come, ye blessed of my Father, and
inherit the Kingdom prepared for you, from the foundation of the world.'" This and other passages of Scripture
my mother taught me before I was really able to pro-
nounce the words after her. All this was my religious
instruction, besides what I learnt in the Sunday school.

CHAPTER II

IN 1834 the curate-in-charge and his sister left our parish, and moved into Berkshire. Before the curate left he came to say good-bye to us. He also brought us some very useful things, which were most acceptable, for I know my mother had to struggle hard against wind and tide, as one might say, to keep us six great rollicking boys tidy, and how she did it as well as she did, with the scanty materials at her command, I really cannot conceive ; but I do know that she many times went without food, so that we might have our fill. The curate looked at my sister's seamed face, then patted the baby, and said, " Surely, Mrs. Langdon, you do not want so lovely a child to be disfigured with small-pox, do you ? "

" What can I do to avoid it ? " asked my mother. " We have always been taught by our clergy that all these evil things are the ' Lord's ' will, so who can hinder it ? "

" God's will ! " answered the curate. " Have you not heard what everybody is talking about, I mean vaccination and cow-pox ? Vaccination is a process by which matter from a cow is inserted into your child's arm, and in the course of a few days the child will have what is called cow-pox ; it is exceedingly mild, and the child will not suffer much, and if properly carried out, it is a sure preventive of small-pox."

"Dear, me," said my mother, "I wish I had known this sooner. I would gladly have had all my children vaccinated."

"I am very glad to hear you say so," replied the curate, "but it was only accidentally that I mentioned it to you. I have spoken to several people about it, and I have found them so thoroughly prejudiced on the subject that I have found it prudent to hold my tongue. Go to Martock to Dr. Stuckey, and I know he will vaccinate the baby free of charge, and as I am leaving the village to-morrow I am very sorry that I shall not be able to know the result of the operation."

My father had in the meantime come home to dinner, and had heard the latter part of the conversation, and he said, "I will take care you know the result, sir. When it is all over I will write and send you a letter, and let you know all about it, and my wife and myself are truly grateful to you for mentioning it."

I have only to add that my father did write the letter to tell the good parson all about the baby having been vaccinated, and he had to pay one shilling and tenpence for that letter to be posted, besides having to walk several miles to the office.

Here I must say a few words in reference to Nanny Holland, and how it was that such an old shrew should be able to hold such power over nearly all the housewives of the village. In her younger days Nanny had the reputation of being exceedingly skilful in midwifery. Moreover Nanny had been known more than once to set a broken leg, or arm, when the doctor was too busy, or, which was often the case, too drunk to attend. Nanny was always ready to assist her neighbours in cases of sickness. She would go when called upon, whether by night or by day ; and if any one hesitated to call her, she would not be any the better pleased, and would give them what she called a bit of the rough side of her tongue.

But there was still another reason, and not an unimportant one, why the women in the village did not always consider it prudent to offend old Nanny. At that time almost, if not quite, every wife in the village made her own bread, and Nanny had the only oven in the parish ; and there the women would go, carrying their dough with them, to be made into loaves and baked.

In those days there was no electric telegraph, but somehow or other news would fly ; and my baby sister had not been vaccinated many hours before the news reached old Nanny's ears, and she took the first opportunity to call and find out particulars, so she came in with her teeth clenched, and her dark eyes sparkling with rage, and said, " I have just heard some news, and what dost thee think 'tis ? Why, I heard that thee's been down to thicky doctor and had thy chiel knockle-headed with vaccination.".

" Well, so I have," replied my mother, " and I am only sorry that I did not know of vaccination sooner, so that I could have had all my children vaccinated."

" Whew," said Nanny, " of all natural fools I ever knowed, thee art the cussedest fool of all. Mind'ee, if thee brings thy dough to my bakehouse Vriday next, I'll kick thee and thy dough out vaster than thee brought it in." And without another word old Nanny went away, and from that day forward she always gave my mother a wide berth. Whenever they met she would cross the road and pass along on the other side.

I have stated that our curate-in-charge had left the village and gone into Berkshire. The rector was a gentleman whom I had never seen. It was reported that he was squandering his time and his health and wealth on the turf, amongst thieves, black-legs, thimble-riggers, and other rogues and vagabonds. I know that it is not always prudent to believe all that is stated by the tongues of the villagers, but in this case I fear the accusation was only too true ; in fact, the probabilities were that in this case the village gossip did not know all the truth. One thing was certain, he had to go about incognito as the bailiffs of the county court were constantly looking after him to serve him with a writ, or to arrest his person. Only one good thing do I know of him ; he used to send four pounds a year to my father for the maintenance of the Sunday school.

After our curate left, it was several months before we had another. The parsons of the neighbouring villages used to come, and sometimes we had morning service, and sometimes afternoon, and sometimes evening service, and more often no service at all. I remember on one occasion the bells were chimed at half-past ten, and the people came to church, but no parson came ; the project again was tried at three, and again the people came, and again no parson, and as a sort of forlorn hope the bells were chimed again at six, and still no parson came.

Old George Pant and a few others set up services about this time in the blacksmith's shop. Now old George Pant wore a wig, and other boys and myself used to go and peep through a large crack in the door of the blacksmith's shop, and watch him while he was praying. He used to get dreadfully excited and shake himself about, till by and by his wig would drop down behind him.

I had seen George Pant shake off his wig more than once, and the wicked thought entered my mind to try and steal that wig, which piece of theft I actually did accomplish on the very next Sunday evening ; and this is the way I did it. The door of the blacksmith's shop in which the meetings were held had in it several large cracks which I could easily put my hand through ; and I noticed that when George was praying, he, and all his congregation, knelt with their backs towards the door ; and so intent were they upon their devotions, that one could open the door and go in and out again, without attracting their attention. But I was too prudent to risk myself in far enough to pick up the coveted wig, when it should chance to fall ; so I provided myself with a long stick, and tied a couple of eel-hooks to one end, and watched my opportunity through the crack in the door. I had not very long to wait. George began to pray, and presently down came the wig. Directly it touched the ground, my fish-hook caught it up, and in another instant I was out of sight of the door with the wig under my arm. But I was no sooner at a safe distance, than I began to reflect, and I would have given the world to restore the wig to its place, but I knew I dared not do it. I knew that if I gave it up to its owner, he would freely forgive me, but my father would literally skin me. So I dug a hole in our orchard at the foot of the quince tree, dropped the wig in, and simply held my tongue. Its sudden disappearance was a nine days' mystery in the village.

Meanwhile my father and mother were doing their level best to keep the Sunday school going with no help from any one ; and there we children were taught the catechism, and lessons from the Old and New Testaments ; and the stick was frequently used. At length we received news that a new parson was coming, and all sorts of speculations were rife as to what sort of parson he would be. Was he young or old ? Married or single ? Rich or poor ? At last the bells were set ringing, and every boy blew his penny whistle and fired his pop-gun, because the Reverend Peter Manonni Scrope Cornwall, M.A., with his two sons, and two daughters, and sister-in-law, Miss Brown, who was his housekeeper, had actually arrived. Mr. Cornwall took the curacy at £80 per year, and an old tumbledown, damp, dismal den of a house to live in. Now the Rev. P. M. S. Cornwall had to preach the Gospel and educate his four children, keep up the dignity of his profession, visit and succour the sick, give to missionaries, buy books for the Sunday school, and subscribe to the thirty-nine articles, besides giving an annual treat to one hundred children, all out of £80 per annum. All of which he did to perfection.

Mr. Cornwall was what may be called a good-natured, good-tempered sort of man ; somewhat inclined to be stout ; and I know that if any one was troubled in mind, body, or estate, they had only to go and open their heart to Mr. Cornwall and they would be sure to find a friend. Moreover he had such a pleasant, benevolent-looking face that those who saw him were bound to love him. His sister-in-law, Miss Brown, was a lady of independent means, and when people went to their pastor to complain Miss Brown would almost surely be present, and she would put in a word here and there, as the case might require. She would blow people up if she discovered that their grief was brought about by their own naughtiness, as she would term it. She would tell them that they must be born again, and that they must go regularly to church, and after she had told them of their faults and how to mend them, she would dip her hand down into the recesses of her great wallet and bring up half-a-crown, and hand it over to the grieved one and say, " Bless your heart, you must give God the glory, you must pray all day long, bless your heart, and say, ' Create in me a clean heart, O God, and renew a right spirit within me,' and then you won't fall into trouble again, bless your heart."

The very first Sunday she was in the village she went and took full possession of the Sunday school, and asked my father to give her the names of all the people who had children but did not send them to school, saying

she would go round and ask the fathers and mothers
to send their children. So during the week Miss Brown
trudged from house to house, and asked the parents why
the children were not sent to school. Most of them
began to make excuses. Some had no clothes fit to go in,
some had no shoes, some were sick with influenza, some
were getting well, and others getting ill with small-pox ;
in fact, some had real excuses, and some made paltry
excuses. But Miss Brown was equal to the occasion.
Those who were ill were to come to the parsonage for
medicine, others were to come for clothes, shoes, or
hats. Anything and everything could be had for the
fetching of it, and it was really astonishing how Miss
Brown came by garments to suit nearly every child in
the village ; if she had been a marine store dealer, she
could not have been possessed of more odds and ends, so
that fathers and mothers as well as their children had
not a shadow of an excuse for not coming to church or
school.

But the funniest thing was that Miss Brown did not let
old George Pant escape her notice. She called upon
him, and his wife began to make excuse for him, that
he had no hair on his head, and that he used to wear a
wig, but some mischievous person had stolen it, and
that George could not go to church and sit in a draught
or he would catch such a cold that he would not get
well again for months. Miss Brown listened as patiently
as she could, and then said, " Bless your heart, I have
a wig that was my uncle's ; and if George Pant will
come or send to the parsonage he shall be most welcome to

it. It will just suit his complexion, bless your heart, and if people will only pray to the Lord, He will always give them what is good for them, bless your heart. You must ' seek the Lord while He may be found, and call upon Him while He is near.' "

The Sunday school was held in the church, there being no school-house in the parish, and every Sunday, in all weathers, at nine in the morning and half-past two in the afternoon, Miss Brown would be at the church door waiting to go in and open school ; and I do believe that Miss Brown's great gold watch was always half an hour too fast, for my father, who was the very cream of punctuality, could not keep time with her. Father kept the keys, and he was not always ready to open the church doors when the time was up by Miss Brown's great gold watch, and when he did arrive she would give him a gentle reminder that he was not in time by pulling out of her great wallet that great gold watch, and saying as she did so, " Come, come, children, in to school ; we are already two minutes late, and we have no time to lose ; come and read." All had to read or learn a text, and were taught the catechism, before the afternoon service, which began at three o'clock. But Miss Brown was very tender-hearted towards her brother-in-law, the curate, and if that gentleman happened to have a cold, or a touch of the gout, which happened very often, then somehow there were fifty or fifty-five minutes between half-past two and three o'clock by Miss Brown's great gold watch, because " bless your heart, Mr. Cornwall has a bad cold and cannot walk very fast."

Also Mr. Cornwall came amongst us most Sunday evenings and gave us some wholesome admonition, and he would tell us of all the most interesting things that were going on in the outer world, and of which we should never have heard without him. And when the dear man stood there in our midst telling us all these stories, his face beaming with goodness and kindness, and his hair as white as snow, I think I almost worshipped him. Then about every sixth or seventh Sunday he would preach a sermon specially to the young. Thus did the Rev. P. M. Cornwall and Miss Brown take possession of the hearts of the people, both old and young, and in a very few years boys and girls grew up, and, as young men and maidens, still attended the Sunday school— a school that could not be matched for miles around. On Easter Monday Mr. Cornwall would invite us to meet him in the churchyard, and we would join hands and encircle the church. Then he would feed us with hot cross buns, and do all in his power, with the help of Miss Brown, to make us happy. It seems needless to record how much these two good people were beloved.

CHAPTER III

STARTING IN LIFE

A T the tender age of eight I was sent to work on a farm belonging to Joseph Greenham. For the princely sum of one shilling a week I had to mind sheep and pull up turnips in all winds and weathers, starting at six o'clock in the morning. Very often I was out in the pouring rain all day and would go home very wet, and then my good mother had something to do to dry, not only my wet clothes, but also those of my four brothers. And I know it took her half the night to mend and tidy all our clothes. As soon as I was able I had to go driving plough, for in those days a man would not think of ploughing without a boy to drive the horses. Now it was my sad fate to be placed under the hands of the most complete vagabond that it was possible for the spirit of all evil to beget. I cannot here tell—and if I could, nobody would credit—the dreadful usage which I received from his hands.

Although Mr. Greenham was my employer, yet to all intents and purposes Jim the ploughman was my master. I was completely in his hands and under his control, and it was in his power to do what he thought fit. There was a public-house in our village kept by a widow, whose name, curiously enough, was Temperance Patch. Jim was one of the best customers that Temperance Patch had. He spent all the money he could earn, beg or steal, in her house, and when he had no cash, he did not scruple to steal his employer's hay, corn, straw, eggs, fowls and potatoes ; in fact everything portable was carried away to the *New Inn*. I once thought it my duty to report to Mr. Greenham that Jim had carried away a large bundle of hay, and when Mr. Greenham taxed him with theft, he cursed and swore, and said that I was a wicked young liar. After this, until I was thirteen years of age, my life was not worth the living ; for I was thrashed and kicked and beaten most unmercifully by this brute. So I learned that a still tongue makes a wise head, and never once again did I say anything to any one, not even my mother, about the cruel treatment which it was my lot to receive. Jim used to make me harness the horses long before I was tall enough to reach their heads, and beat and kick me if I could not do it quickly enough for his liking ; and I used to wonder every day and all day, and ask myself, " Why was I born ? " Sometimes Jim would lie down under the hedge and go to sleep, making me plough the ground the while ; and although I was but a child and scarcely tall enough to reach the plough handles, yet if he woke up and found any bad ploughing he would beat me to his heart's content. But with it all, he never could get me to tell the abominable lies that he would put in my mouth to tell Mr. Greenham so as to save him a scolding when he had been neglecting his work. I had learnt from Mr. Cornwall and also from my father that lying lips are an abomination to the Lord, and this feeling was so strong within me that I could never corrupt my conscience and degrade myself to repeat Jim's falsehoods, and I came in for many castigations accordingly.

On one occasion Mr. Greenham took Jim and me to the cellar, to lime some wheat before it was sown. While we were there Mr. Greenham was called away, and directly his back was turned Jim caught up a dipper, as if he had not another moment to live, drew some cider and drank it greedily down ; then he drew some more and offered it to me, but I refused. With an oath he pressed the edge of the dipper against my lips until they bled with the pressure ; at the same time he held me by my hair, in order as he thought to pour the stolen liquid down my throat ; but Jim did not succeed in his purpose, so he drank it himself and threatened, using fearful impreca-tions, that if I ever said a word about it he would kill me on the spot. I don't think I should ever have said any-thing about it, but thieves are generally great fools. Jim in his greedy haste did not turn the tap back as it was before, so that there were a few drops on the pavements. The dipper, also, was wet and smelled of cider. So Mr. Greenham accused him, but Jim began to call God to witness that he was as innocent as a dove, and he had the impudence to refer to me to prove his honesty. The master asked me and I told him the simple truth, knowing full well that I should catch it soon. As soon as Jim's guilt was discovered beyond dispute, he began to shed crocodile tears, and to lament and beg pardon in such a humble and seemingly contrite manner that the master's eyes were blinded, and he forgave him there and then.

The next day we went into a field to plough, and now my punishment began. Jim belaboured me with the horsewhip as long as he felt disposed. He knocked me down and tried to jerk the breath out of my body. Then he wrenched my mouth open with a large nail and filled it with dirt. He allowed me to get on my legs again and resume ploughing for a time, but he soon began on me again. He struck me down and kicked me, and danced upon me, till I felt very faint and ill with loss of blood. I really thought my end had come, and I felt very glad. It may seem rather paradoxical, but that moment was the happiest moment of my life. I thought of dear Miss Brown, and her teachings : " Blessed are they that mourn, for they shall be comforted. Blessed are the pure in heart, for they shall see God. Blessed are they which are persecuted for righteousness' sake, for theirs is the kingdom of Heaven." All these and other precepts flashed into my mind, for I knew it was out of envy that I was so cruelly used. But somehow I refused to die at his bidding, so Jim waited for another time to try and send me out of the world as if by accident.

One of the horses was exceedingly ticklish when touched in a certain way upon its backbone, and could not bear to be touched on this particular spot with a curry comb, and sometimes when so irritated would let fly with both heels at once. So on the morning following the last punishment Jim set me to clean some portion of the harness, and made me stand in a certain position directly behind the ticklish horse. There I worked away without any idea that mischief was brewing. Jim, however, had laid all his plans, and if they had succeeded and I had been killed, he would have been found blameless. There was an open window to the stable exactly opposite and close to the ticklish horse, so that a man outside, by standing on a ledge of the wall, could put his hand through and touch the horse's back. I heard the horse make a noise, and on looking up saw Jim's head outside the window, and his hand upon the horse's back. At the same moment the horse let fly, and one of his heels came against my left side and sent me dashing against the wall. I knew no more until I found myself in bed with my mother crying and washing the blood from my hair and face, and felt a great pain in my hip, where the horse's hoof struck. There was also a big scar on my head where I was knocked against the wall. I can only account for not having been finished off that time by the fact that the horse did not kick when it was first touched, but began to prance about, which arrested my attention and I moved close to his heels. If I had been a little further off his heels would have struck my head or the upper part of my body and I should not have been here to write.

After lying in bed about a week, where I cogitated and wondered for what earthly purpose I was born, I had to go back under this fiend again. Every other place in the parish was filled and my parents could not afford to keep me in idleness, so there was nothing for it, but to go back to work again as soon as possible.

A few days after this the very same horse got restive in a field where we were and turned over a cartload of manure upon poor Jim. I thought he was killed, in fact for a moment I hoped he was killed. But immediately I would have given worlds to have called back the thought. Miss Brown's words came upon me, quick as a lightning flash, " Create in me a clean heart, O God, and renew a right spirit within me." Other of her precepts came strongly into my mind, and I shook with fear, for I had learned that to wish a man dead amounted to the same thing as killing him. Therefore, I felt that I had committed a most grievous sin, and I cannot express the joy I felt when I saw Jim crawl out from under the cart unhurt. He began to curse and swear at the horse and me, saying it was all my fault, whereas it was his own fault, as in harnessing the horse he had negligently left the buckle of a strap under the cartsaddle, so that the buckle rested exactly upon the backbone of the horse and caused him to be restive.

I was under Jim's control for five years—years of my childhood, which I ought to be able to say were the happiest of my life. But they were just the reverse, and if I stated all that I suffered at his hands, no sane person would believe that such things could have been done with impunity.

Not many years ago Mrs. Beecher Stowe shocked the refined feelings of the civilized world with her graphic account of the sufferings of the negro slaves in the United States of America. I cannot write my history in the shape and manner of a novel, with its parts and counter-parts, but what I have written are some of the main facts and features of my boyhood life. Some people, those who have passed smoothly through their childhood, and have scarcely known sorrow, may ask whether it is possible that such things could have been done in England ? My answer to this is, yes. It was not the parents, but the age that was to blame, as may be learnt from some of the works of Charles Dickens, and other writers who have given pictures of the period. I know that my brothers could write a parallel history, and they were not under the hands of so complete a blackguard as it fell to my lot to be under.

When the season for ploughing was over I used to get a few weeks' relief from the hands of my tormentor. During such times I was sent into the fields minding sheep. These were days of pleasure and happiness. I had to work hard, but toil was a pleasure as long as I had no one to abuse and ill-use me. I was the happy possessor of a tattered Testament, and I used to read from its torn pages. It began at the words, " Let not your heart be troubled," and ended with the twenty-seventh chapter of the Acts. I read the first and last chapters more than all the rest, and really knew them all, every word.

Now dear old Mr. Cornwall used to come out in the fields and find me out and ask me questions about Scripture history, and I believe I used to answer him to his satisfaction for he called me a good boy. As far as I know it was the first time I had been called a good boy except by my mother, and I fancy I grew an inch taller all at once and that his calling me a good boy had a very

strong influence in making me try to be good ; but whenever he talked to me my conscience pricked me relative
to old George Pant's wig. I never could forgive myself
for stealing it, and would have confessed to Mr. Cornwall
concerning it, but I thought he would tell my father, and
I did not want an extra thrashing.

I used to leave work at six o'clock, and Mr. Cornwall
told me that if I would come to the parsonage and pull
up the weeds in his garden path he would give me a shilling. The idea of having a whole shilling, all in a lump,
frightened me. I had never possessed a coin of the realm
above the value of a halfpenny, and such halfpennies
were, like angels' visits, few and far between, for the wages
which I earned had to go for my maintainance. So I
went every evening to accomplish the work, and was very
particular to do it well, so that nothing should prevent the
free and unconditional receipt of the shilling. I had
been to Crewkerne a few weeks previously and had seen
a book in the printseller's window ; it was *Pinnock's
Catechism on Astronomy*. My heart had been aching to
obtain that book, but the price was ninepence, and I
knew that if I saved up those very scarce halfpennies it
would be years before I got ninepence, and so I thought I
should never get the book. But now a new light had
unexpectedly fallen upon the subject. My dream of
possessing "Pinnock" would now be realized, and that
much sooner than I ever had imagined. I should now be
able to run over to Crewkerne and buy the book and have
threepence change. Therefore I finished my task, and
swept and cleaned up all the weeds, and with a joyous

heart I presented myself at the parsonage door for my promised shilling. "Put not thy trust in Princes" is a trite saying, but oh, how deeply and grievously I realized its truth, for I never received that shilling. Mr. Cornwall was laid up with a fit of gout, and what with the twinges of the malady and the business of his curacy, I suppose he had forgotten me.

When Mr. Cornwall was upon his feet again I was too shy to ask him for the shilling and so it was passed by, and I was compelled to go without the pleasure of reading "Pinnock" for several years. About two years after this Mr. Cornwall came out to see me in the fields. I had gone to another field a mile away, but had left my jacket and some tools and my fragment of a Testament, all rolled up together in a corner of the hedge which I had been in the habit of using as a dining-room. So the parson thought he would be inquisitive. He opened my jacket and found an assortment of things that I had cut out of sticks and turnips. There were ships, soldiers, sailors, anything and everything, and I afterwards heard him tell father that he had fairly roared with laughter on finding them.

A few days after this I saw the dear old face coming up the side of the hill where I was with the sheep. He was approaching very slowly; he never could walk very fast across the fields, because Miss Brown always would insist on his wearing a pair of her old clogs that he shouldn't catch cold, "bless your heart." When he came up to me he began to ask me questions and whether I found time to read the Bible. So by degrees he got me to show

him my fragment of a Testament. He turned over the
leaves and returned it to me. Then he pulled out from
his pocket a brand new Bible. It was a reference
Bible, such a book as could not be bought at that time for
less than seven or eight shillings. Mr. Cornwall gave it
to me and told me to read, mark, learn, etc. In the fly
leaf he had written :—" Presented to Roger Langdon,
for his good conduct at the Sunday school, by the Rev.
P. M. S. Cornwall."

I cannot describe my feelings on that occasion. I believe I laughed and cried. I kept that Bible, and carried
it about with me wherever I went ; until, a few days before
I was married, it was stolen from my lodgings in Bristol.

CHAPTER IV

SINCE Jim had compelled me to plough the ground while he slept, or otherwise idled his time, by the time I was twelve years of age I could plough a straight furrow. It was considered a crime of the deepest dye to plough a crooked one. There was a ploughing match to come off at Haselbury, and Mr. Greenham entered me on the list as a first-class boy, and Jim was entered as a first-class man; we then had to practise side by side in a field of clover. Everybody said I should win the head prize, which for the boys was £3, while the head prize for men was £5 with a society's coat and buttons. At length the important day arrived, and everybody went to the field where a piece of land or ridge must be ploughed in fourteen rounds; that is to say, the plough must go across the field and back again fourteen times, which makes twenty-eight furrows for each piece of land. Neither more nor less must be ploughed in these twenty-eight furrows. All the furrows must be straight, no grass or weeds must be seen sticking up between them, and when finished the ridge must be level and even. Now there was one ridge which had a hollow across it in a diagonal direction, and as all the pieces of land to be ploughed were drawn for by lot, this piece came to my share.

I asked the umpires if they expected me to get that hollow up level, and they said Yes, most decidedly; but they would give me an extra half hour to do it in. So at it we went, whistling the tune of " God speed the plough and the Devil take the farmer." Now, notwithstanding the hollows, I had the pleasure of hearing my work praised exceedingly, which gave me much courage. I finished my piece as soon as the rest had finished theirs. We then had the pleasure, the inestimable pleasure, of seeing the farmers and gentry eating roast reef and plum pudding in a decorated barn. The smell of these good things ought to have done us good. For myself I was very hungry, having had nothing to eat since before six o'clock in the morning. When they had dined we were called in one by one to receive our rewards. I heard my name called, and went into the barn to receive my prize. The squire, who was the spokesman, praised my work and said that I should make an excellent plough-man, and had it not been for the hollow, which I had not fetched up to their satisfaction, I should have been entitled to the first prize. But as it was I should be awarded the third prize of £1.

Now Jim got nothing. I do not know why, for he was undoubtedly a good ploughman when he chose to put himself into his work. So he sent me home with the horses, as it was getting late, and said he would go and get my £1. I had no choice but to obey his orders,

and I never saw my money, for he went straightway to
Temperance Patch's, and spent it there. My poor mother
went to him about it, but she received nothing but
curses.

So now I began seriously to think how I could get out
of his power. I used to measure myself once a week to
see how soon I should be tall enough for the Army. I
was thirteen years of age, and big and strong. So when
Her Majesty Queen Victoria was crowned I went to
Yeovil, and for the first time in my life saw and heard a
military band. I asked them to take me, as I could play
a flute or cornet, but they replied, " You must grow a
bit, and then call on us again."

I grew a good deal during the next year, when I was
fourteen. I was determined to bear Jim's cruelty no
longer, and I knew I was now tall enough for the Army.
So I made preparations for a start. I was receiving four
shillings and sixpence a week, which was good wages at
that time ; but of course this all went for my maintenance.
I went to my master, however, and asked him to keep my
wages for three or four weeks, so that I might take it all
in a lump, but I shrewdly held my tongue as to why.
It was the only occasion on which I did not act quite
openly with my mother, but in my mind I knew I should
make it up to her later. So when my wages amounted
to nearly £1, I asked Mr. Greenham for it and then
made a start. I put together what few clothes I had and got
up at three o'clock in the morning, with mixed feelings
of joy and sorrow secretly bidding adieu to mother,
father, brothers and sisters.

On arriving at the end of the village I glanced back to take a last look at the church and steeple, which were just discernible in the grey dawn. I thought of Mr. Cornwall, and Miss Brown, some of whose teaching again came into my mind. I remembered that I had so often repeated the words, " The Lord is my Shepherd, I shall not want," and " I will lay me down in peace and take my rest, for it is Thou, O Lord, that makest me to dwell in safety." I now began to make tracks for Weymouth, thirty-three miles distant, and beyond the first seven miles I did not know an inch of the way.

I trudged on, however, like a snail, carrying all I possessed on my back. After about nine or ten hours' sharp walking I arrived at the quaint old town of Dorchester, where I walked leisurely through the streets, looking at the old-fashioned houses, and such things as took my attention. Just as I was going to turn into a shop to get some bread and cheese, a smart recruiting sergeant came striding up the street towards me. He clapped his hand upon my shoulder and said," Here, young man, will you enlist ? " I do not know whether the perversity I showed is a characteristic of human nature, or whether it belonged to me individually, but if he had asked me to take a dose of poison, I could not have felt more vexed and annoyed; and when he showed me the shilling, the disgust I felt was beyond description. Perhaps the fact of my being very hungry had something to do with it, but at any rate the idea of being a soldier went entirely out of my head for ever.

After an hour's rest in Dorchester, I travelled forward

I still had about eight miles to go, and my feet were already blistered. I scarcely felt it, however, as I had often had them blistered when ploughing, with Jim's ill-usage into the bargain ; but now I had freed myself for ever from his cruelty, and I went along with a light heart.

I reached Weymouth about five o'clock in the afternoon. I had done the thirty-three miles, including stoppages, in fourteen hours. I walked round the harbour and asked every skipper I could find if he would take me on board his vessel in any capacity, but from all came the same answer, " No." So I began to think that the world and its inhabitants were not exactly what I had always thought or fancied they were. All the sea-songs I had heard, " The poor Sailor Boy," " The Cabin Boy," and others, had led me to believe that the skippers of vessels were only too glad to get hold of a boy when the chance offered ; but now I found out my mistake. But one man told me that a shipowner in Jersey was in want of hands. I made inquiries and soon found out that a steamer would start for Jersey at nine o'clock the same evening.

I went on board, paid ten shillings for my passage, and was soon off towards Jersey.

I cannot describe how much I enjoyed the view of the sea, especially when we began to lose sight of Portland. The moon was shining and I could look around and see a great expanse of water. The sea was not rough, although there was a swell which was sufficient to toss the vessel up and down in what I soon found to be a very disagreeable manner. I soon began in right earnest to feel what the French call "mal de mer."

After a twelve-hours' passage, which was considered good work at that time, we landed in Jersey. I at once found my way to the shipowner, and he set me to work there and then unloading salt from a ship at one shilling and sixpence a day. The salt was not in lumps, but in the form of small grain. It was at that time used in England for manure ; but in Jersey and France it was used for salting butter, meat and fish. To any whose shoes have been worn off their feet, and their feet blistered and made sore with walking, I need hardly say that the salt soon found its way into my boots and made me nearly faint with pain. However, it was soon over, and in a couple of days I believe my feet were hard enough to walk upon flints without any inconvenience.

Sunday soon came round, and I brushed myself up and went to St. James' Church. As I glanced over my person I found that my only suit was very much the worse for wear after unloading, but I had no idea of staying away from church on this account. In the afternoon I went to the town church and saw the soldiers walk down from Fort Regent, with their fife and drum band playing " The girl I left behind me " right down to the church door. As soon as they were out again the band struck up " The Irish Washerwoman." I thought it was very wicked, and wondered what my father would have said about it. On the Monday morning, the judge of the island, who was my employer, came into the stores, where I was at work, racking brandy, and adding water and logwood and a drug, the name of which I do not remember, by which means we made three hogsheads out of two. Now the judge was one of the richest merchants in the island, and therefore thought it his duty to set a good example to the other merchants, and to all seafaring men. He was a large importer of wine, rum, brandy, and gin, and was always very particular to see that we did not put too much water and logwood into the brandy, and water and sulphuric acid into the gin. So that his spirits were always considered the best in the island. The judge went regularly to church, and as he was walking round the stores on this particular Monday morning he came up to me and said, " Young man, I was much pleased to see you at church yesterday. Are those the only clothes you have ? " I confess that these remarks about my clothes did not please me, because on the Sunday I had gone to church feeling unhappy at not being able to change my garments, and hoped that no one would notice me ; and when I found that the judge had noticed me, I must say I felt seriously annoyed. I answered him civilly, however, that it was my only suit, but that I should get a new one at the first opportunity. The judge raised my wages there and then to two shilings a day, and in the afternoon called me into his private office and told me to go to his house as his wife would like to see me.

During the rest of the day I was nearly beside myself, wondering what on earth such a great lady could want with a poor boy like me ; therefore I was twisting and turning the thing over in my mind, wondering whether she could

speak English, and what she would say to me and how
I should answer her. At length evening came, and I
went to the house and was presented to her. She was
exceedingly plain in her dress, as all Jersey ladies were,
and I could scarcely believe it was the wife of the judge.
However, she soon told me what she wanted me for. She
said, " J'ai some clothes for de garçon ; de coat was too
much big, go to de tailor and have one gros cossack too
make little from fit for yourself. Reste vous one little
minute, je donne vous some-sing for to mange." The
drift of all this was that she gave me a meal and a
left-off suit of the judge's. He was one of the largest
men in the island. I was a mere stripling, and I believe
I could have stood in one of the legs of the trousers.
I could not get a tailor to do anything with such a suit
unless I paid him more than a new suit of clothes, and I
did not go to church for several Sundays, until I had
saved enough to buy myself a decent suit.

Now the judge was known to be a very pious man.
He would take hold of little dirty urchins in the street
when he heard them using bad language and reprimand
them. His own people had to pull a long face and assume
a virtue if they had it not. And when the postman
came round with a petition to get off his Sunday duty,
everybody in the district signed it except the judge.
He could not forego the pleasure of receiving and reading
a few letters and allow the poor postman to go free on
a Sunday. I have been greatly amused more than once
on going into an Assize Hall and seeing the judge sitting
in his chair, looking as grave and solemn as only a

judge can look. The clerk of the court read the Queen's proclamation against vice and immorality ; solemnly called upon the magistrates and sheriffs of counties to use all their power to suppress all kinds of vice and lewdness, especially Sabbath breaking ; and yet the judge could not allow the double rap at his door to cease on a Sunday.

I got on capitally with this lady ; we seemed to understand each other at first sight. But do what I would, I could not bring myself to feel the respect due to her, simply because she wore a pair of old and dirty wooden shoes, a short, rough woollen skirt, a great red-patterned kerchief over her shoulders, and a large, stiff, white muslin cap on her head. Altogether she cut such a figure, that I could not fancy she was the wife of a rich merchant and judge. But I found after a while that the ladies of Jersey were exceedingly plain and unassuming. They assisted in house and dairy work ; they milked and fed the cows. It was a very common thing to see the farmers' wives and daughters milk cows into one can, and goats into another ; then, tying the cans together and slinging them across an old horse's back, they would perch themselves on the top, and set off to town at five o'clock in the morning, to sell the milk from door to door. They returned to breakfast and spent the remainder of the day working in the fields. I saw them, both in Jersey and in France, actually ploughing, sowing, reaping and mowing ; and yet these people were rich and had their thousands in the bank. After witnessing how hard the women had to work in Jersey and France I was not surprised that Napoleon I. said that England was a paradise for women.

I continued in the employ of the judge all the summer My usual work was to adulterate the wine, brandy, gin, rum, and whisky; and though constantly amongst this firewater, I am thankful to say I did not acquire the taste for any of it. Yet all who worked there could have what they liked. The judge gave carte blanche. I often thought what a paradise this would have been for Jim, how he would have made himself a perfect walking swill-t ub; but it would have soon killed him. I watched many strong sturdy fellows from Devon and Cornwall actually kill themselves with the accursed stuff. Not that they were drunkards; nothing of the sort. But because brandy could be purchased at sixpence a bottle, so they would constantly be sipping it. They did not get drunk, but would take a little in winter to keep the cold out, and a little more in summer to keep out the heat; they would soon get " brain fever," or as some people would say, sunstroke, and die ramping mad. I have seen and known this in many cases both of men and women.

CHAPTER V

THE judge had a fleet of ships of his own trading to nearly every corner of the globe, and in the months of September and October several vessels returned from the Newfoundland fishery, laden with codfish, whalebone, sperm oil, and seal, beaver, fox, and other skins. He made me a sort of deputy-clerk, and I had to note down every article with its number and weight. This I did so much to his satisfaction that at Christmas he actually gave me a sovereign as a present over and above my wages; and a few kind and complimentary words that he spoke made me feel as if I had suddenly grown an inch taller. Now I began to feel very pleased and glad I was born.

I began to think myself surpassingly rich, for I had three good suits of clothes, and six golden sovereigns in my pocket, and I thought of poor Jim the ploughman, who used to go to bed early on Saturday night to give his mother the opportunity to wash his one and only ragged shirt, so that he might have it clean on Sunday morning. All this time I had thought greatly about my poor mother, for she had not the slightest idea of what had become of me. I had been away from April until Christmas.

I had never up to that time any idea of writing a letter as I had never been to a week-day school, but somehow or other I had acquired the art of writing. Of course I could not realize how very deep her grief must have been, being only a male. We as men love our children, but our love at best cannot be measured or weighed against the deep and constant love of a mother for her child. Now I thought how pleased she would be if I could only just put these six sovereigns into her hands. I thought whether I could manage to go for a week to see her, and return again ; but on second thoughts I decided that would not do at all. So at last I wrote a letter, and told my mother that I was well and flourishing, but could not come to see her. If she would come and see me, however. I would send her the funds to pay her passage. She answered my letter by return of post to say that she would come, and so in the course of a few weeks I had the pleasure of meeting her on the pier, as she landed from the *Atlantic.* The judge gave me as many holidays as I wanted, and so I was able to show my mother about the Island.

The first thing that surprised her was to see the French women wheeling heavy barrows of luggage about, while their lords and masters were swelling about with their gold-laced caps, gingham blouses, patent leather boots, and the everlasting pipe in their mouths.

Mother would gaze at the women in the street, and say, " Well, I have worked hard in my time, but I am very glad I am not a Frenchwoman." She was not surprised, she said, that the Duke of Wellington was able with a

D

handful of Englishmen to go over and thrash them French-
men on their own ground.

I thoroughly enjoyed myself during the five weeks
mother stayed with me. Father had given her four
weeks, but I must confess to playing a trick on her so
that she could stay with me another week. On the day
that she was to leave I went down to the steamship office
to ascertain at what time the packet started, and found
it was timed to start at nine o'clock in the evening.
But my mother did not trust quite to that, for she went
herself to learn the hour of departure, so it was no fault
of her own that she was left behind. Well, we had high tea
at six o'clock, and got a few friends together just to
talk away the hours until nine o'clock. So while the
ladies were thus employed, and helping my mother put
her things together, I put the clock back three-quarters
of an hour. No one noticed it, as the clock was in another
room, and it turned out exactly as I desired. We went
off, my mother and I, she shaking hands and bidding her
friends good-bye and I laughing up my sleeve, knowing
full well that the steam-packet would have left. When
we arrived at the pier and she found it had gone without
her she could scarcely believe her own senses. I knew
that she had never been away from her home for twenty-
four hours before, and I also knew that my father would
be dreadfully vexed, especially as he would have to
walk a long way to the Foxwell turnpike to meet the
Weymouth coach. I ought to have been put in the
stocks for practising such a joke, but somehow I felt it
might be years before I should see her again.

I took her over to Elizabeth Castle, and she stood close by the great gun when it was fired off, bearing the deafening noise and the vibration under our feet like a real old soldier ; and then we went to Gorée Castle and saw the oyster fishing ; and she was very much amused to see the oysters as they lay in heaps upon their shells, and to watch the mice, that swarmed around trying to get a nibble from the open oysters, which would close up their shells upon them. Then I took her down to St. Clement's Bay, where her grandfather, a sergeant-major in one of the English regiments, landed when he was sent to the island to clear out the French. At the death of Major Peirson in the Royal Square the command fell upon my great-grandfather, who drove the French out of the town at the point of the bayonet literally into the sea, and at the time of our visit the guns which were used in that fight were stuck into the ground along the beach as a memorial of the occasion.

At the time of my sojourn in Jersey from 1839 to 1847 it contained twelve parishes. The capital town of St. Heliers contained six or seven churches besides two Roman Catholic chapels, and several Dissenting places of worship. There was a fine theatre, and a Court-house in the Royal Square. This house did duty as the House of Commons, Guildhall, assize hall, and I know not what besides. I have seen great doings there when a new judge was being elected. I have also seen prisoners tried for various offences, but whether the prisoners were French or English or of any other nation, the whole of the business was carried on in the French language.

If the prisoner at the Bar did not happen to understand that language so much the worse for him.   There was no such person as an interpreter, and I often heard sentence passed upon a prisoner who was quite ignorant of the nature of the trial or sentence until some kind friend who could speak both languages would tell him what he was to expect.

Mr. Charles Carus Wilson, a man over seven feet in height and a member of the English Bar, on one occasion stood up and told the judge that the prisoner had not had a fair trial, that he protested against it, and that he would report the circumstances to Lord Denman, the Lord Chief Justice.   The judge thereupon told Mr. Wilson that he had insulted the court and must pay a penalty of £10, and apologise to the court for such an insult.   " Indeed, I shall do neither one nor the other," replied Mr. Wilson.   " Then," said the judge, " you must go to prison during Her Majesty's pleasure." " Very well," replied Mr. Wilson, " here's off to jail."   So he walked through the streets in charge of a constable, his head and shoulders towering above the heads of the crowd which had gathered round.   In prison they had to put two bedsteads and beds together to make it long enough for him to lie down.

Mr. Wilson, however, took it very quietly and court-eously and reported the whole matter to Lord Denman, who sent over a writ of habeas corpus.   Of course I wondered whatever that could be, but the steam packet arrived on a Sunday morning, covered with flags and banners, and thousands of people went down to see

the sight and wondered what was going to happen next.
I do not know if the judges knew the meaning of it, but
they were nearly frightened out of their wits. Messengers
were sent all over the island to call all the judges together.
On Monday morning they met and consulted, and the
result of their deliberations was that they went themselves and opened the prison doors and asked Mr. Wilson
if he would please to walk out. Charles Carus Wilson,
however, did not please to walk out. He merely replied,
" You have sent me here for I know not what, and I do
not feel disposed to be sent to prison and taken out again
just as it suits your whims."

So the upshot of it all was that they had to pay Mr.
Wilson's fare and their own to London, and all had to
appear before a judge of the Queen's Bench, and the
Jersey judges were fined £100 each, and the poor woman,
whose trial and sentence of seven years' transporta-
tion for stealing a hen and chicken had caused all the
trouble, was freed.

My employer had often told me that if I had been
a few years older, he would have sent me to Newfoundland to superintend his business there. As I
was too young to fill such a responsible position, he
proposed that I should join the *Anchor*, a fine bark of 600
tons. The captain, he said, was a " very nice gentleman,"
and on that vessel I should have an opportunity of
learning the art of navigation, so that eventually I
should be able to take charge of any ship belonging to
the merchant service. I thought this was exceedingly
kind, especially as he said he would provide me with

an outfit, and I then and there closed with the bargain. The *Anchor* was in the harbour and I went on board and assisted in putting in a stock of provisions, ready for the voyage to the Brazils. She was to sail in a fortnight, and I was rather glad that I was born, to fall in luck's way in the manner I had. There now appeared a prospect of my being placed in a position worth struggling for, which I knew was not usually the lot of one such as I. So I looked forward daily and hourly for the kind-hearted judge to supply me with the outfit. When, lo and behold, one morning, a day or two before the *Anchor* was ready to start, the judge told me that he intended to place Jim Drake in my place, because his father was dead, and he was a poor, friendless lad. "Just as if I were not a poor, friendless lad," thought I. I don't think I wished Jim Drake dead, but I did wish that he had never been born. That he should step in and just open his mouth and catch the blessing that was intended for me was almost more than I could bear.

So I had the mortification of seeing Jim Drake go off in the *Anchor*, and I felt more disgusted than I can tell, especially as the skipper was a " very nice gentleman." But I afterwards found out that after all this happened for the best. The judge offered me more than one good situation. He had several other vessels, besides the *Anchor*, but to all his offers I turned a deaf ear. If he had ill-used me, or kicked me and boxed my ears, I should have forgiven him; but he had deceived me, and for that offence I could not respect him. So I left him and sought other employment.

I found work in a large blacksmith's shop, and here I had to work very hard indeed. I stayed for a time, and then engaged myself to another merchant. I went on board his vessel, which was a small sloop trading between France and the Channel Islands and occasionally visiting some English port, Plymouth, Poole or Southampton. Now this old merchant used to go over to Havre or Granville, proceed a few miles into the country, buy cows at from £5 to £8 each, and send them on board the *Medora*, ordering them to be taken to St. Aubin, Jersey, while he himself went on before in the steam packet.

The *Medora* in due course ran into harbour, and there would be the old sinner waiting to order the cows to Poole, and at the next tide we would set sail for Poole, where he again would be waiting to meet us, and the cattle would be unloaded and he would take them to the market or drive them round to the farmers and sell them for pure Jersey cows, thereby gaining an enormous profit.

One Saturday afternoon when the wind was blowing a hurricane and the sea rolling and seething in all its majestic fury, as we were scudding along somewhere between Guernsey and the Isle of Wight, we saw a fine bark in the distance, and on nearer approach we found it was the *Anchor*, returning from the West Indies laden with sugar. We got near enough to speak each other, and I looked to see if I could make out Jim Drake, but I failed to do so. The vessel had encountered some very bad weather for nearly all the bulwarks were gone, and almost everything that is usually carried on deck had been washed away. The men were, like ourselves, tied fast with a rope's end to prevent their being washed overboard. After two or three days and nights in the Channel, we at last found ourselves in Weymouth harbour. We had been soaking wet and had had no sleep or food, nor did we seem to require any, but I went to an inn and got to bed and slept for ten hours. On arriving back in Jersey the first thing I heard was that Jim Drake, who had just returned in the *Anchor,* had summoned his skipper for cruelly ill-using him while at sea. It appeared that the *Anchor* was no sooner out of Jersey than poor Jim Drake began to be sea-sick. So this skipper, the " very nice gentleman," took a rope, and the chief mate took another, and between them they belaboured poor Jim, one hitting on one side and one on the other. So this nice pair seem to have taken a delight in ill-using the poor lad during the whole of the voyage.

One day Jim committed the awful crime of whistling. No one was allowed to whistle on board except the commander, and he only under extraordinary circumstances, such as when there was a dead calm he might whistle for a breeze. Superstition ran so high on board that if a man whistled he was considered an evil genius ; storm and tempest, fire and famine, aye, the devil himself would visit that ship, and from the moment Jim Drake whistled every man on board became his enemy. The skipper cursed, and the mate swore, and poor Jim cried for mercy and said he did not know he was committing an offence. But they tied him to the mast and both took a rope and hammered away as if the subject they were operating upon had been a piece of cast-iron or a block of granite ; and at last one took a handspike and gave Jim a blow which broke his arm, and as there was no surgeon the limb was never set. This was the complaint brought before the court, and when I saw Jim there I cried for pity as the poor lad stood there with his hand twisted round, the back being towards his thigh, the palm outwards, and the whole dangling useless by his side. Now Jim was an English lad, and the skipper a Jerseyman, and all the business of the court was carried on in French. Jim stated his evidence, but there was not a man amongst that crew who would corroborate it, and the skipper and the mate were two very respectable men. Moreover the skipper was such a kind-hearted gentleman that he had actually given his men a double allowance of grog on their homeward voyage ; so of course, as in duty bound, they all to a man spoke of his kind and generous conduct. So poor Jim's complaint fell to the ground. And not only that, but the skipper took proceedings against him for trying to defame his fair character, and poor Jim was sent to prison for three weeks. I cried with anguish for Jim, yet thanked my lucky stars that I did not go on board the *Anchor*.

CHAPTER VI

I HAD now been in Jersey eight years and things began to get rather dead there. Work was scarce and wages very low, and if any one wanted a small job done, there were always about a dozen men ready to do it for almost nothing ; and what made it so was the number of Irish pensioners. They had their pensions, and of course they would do any kind of work for less wages than any other man who had to live entirely by his labour. In fact, things were in a state of stagnation everywhere. I went across to France, and the people there were quarrelling with King Louis Philippe and casting all the blame of their poverty upon his shoulders, and saying how much better they should get on if they could only have a Napoleon to rule over them. They were chalking *Vive Napoléon* upon the pavements and walls.

I went back to England and found things were not much better there. Thousands of poor people were half-starved and half-clothed, and when they asked for work and wages to buy bread for themselves and their little ones, the Commander-in-Chief was ready to fire a volley of grape shot down the street. England and some of the continental nations were at very low water in 1846–7, and I think at their very lowest ebb in 1848. Some monarchs were obliged to abdicate, and in London in April of that year riot and rebellion were rampant. Several thousand people paraded the streets starving. In the House of Commons some one declared that it would be a wholesome proceeding to hang a few rebels. Up jumped Fergus O'Connor, and cried, " Whenever you hang a rebel you should make a point of hanging a tyrant too, and rebellion would soon die a natural death." Soldiers were placed in every town in England, lest the owners of hungry stomachs should show too much anxiety to fill them with bread. At the same time it was said that millions of quarters of foreign grain were brought within sight of the shores of England, but were thrown into the sea by order of rich merchants rather than that corn should be brought in to reduce the price. Such is the greed of man.

The only work that I could get after I left Jersey was in a canvas manufactory in Somersetshire at eight shillings a week, and this was considered very good wages then. Here I found poverty and wretchedness and oppression supreme. There were about four hundred people employed in the various departments of this business ; old men and women, nearing four score, and little boys and girls from five years of age and upwards. Some worked in the factory, and some who had hand-looms took their work to their homes. Several of them had to walk four miles carrying their woof with them, where they had handlooms ; and so these poor toilers worked from early dawn until ten o'clock at night weaving their woof into sail-cloth or canvas. When these poor slaves came to the factory for their work, the foreman would weigh out to each person so many pounds of chain and as many of woof to each person. The weaver had to take it home on his back and produce exactly forty-six yards length of canvas two feet wide. Now if he happened to put a little too much energy or muscular strength into the work, all the woof would be used up before forty-six yards were made. Consequently he would have to send to the factory for a few ounces more woof to finish the pieces, and the cloth therefore would be a trifle too stout, and consequently the weaver would be fined to the tune of any amount, according to the greed and temper of the master, from sixpence up to five shillings and sixpence, which was the wage due for weaving the whole piece. On the other hand if the weaver happened to be weakly and unable to use the strength required, he would come to his forty-six yards' length before he had shot in all his woof. The cloth therefore would not be quite stout enough, and the poor weaver would be fined. Any frivolous pretext was resorted to to fine the workers.

Many a time have I seen poor men or women after toiling hard all the week coming to the pay office for their wages, but instead of receiving any being cursed at and told that it was a very great favour on their employers' part to give them work at all. And so these poor slaves would have to do without the "weaver's ox" (red herring) for their Sunday dinner. A red herring was the greatest luxury these poor people could indulge in, and thrice blest was he who could afford three red herrings a week. There were many other pretexts for fining the weaver besides those mentioned. Another was what was called in weaving a "gout"; that is, in the course of weaving there were some thick and gouty parts in the woof, where the thread was twice or thrice as thick as it should be. If the weaver was not careful, and allowed one of these thick gouty threads into the texture of the cloth, he was heavily fined. The cloth was also examined under a magnifying, glass, and if found pin-holey or spotted, the weaver was fined. All these were certainly very superficial excuses for fining the people, for the masters themselves would throw the canvas down on the floor and walk over it. The cottages where these poor people had to live and do this weaving were shocking hovels.

This firm was carried on by a man and his three sons. Each of the sons took up a certain department. One superintended the spinning department, another the bleaching, and the third the weaving. On a certain Whit Monday at the village club festival an old farmer, in an after-dinner speech, took the liberty to tell the three young men that it was very shabby and mean of them to fine these poor weavers in the manner they did, and he thought it amounted to little less than downright robbery. Then the young man from the spinning depart-

ment jumped up, and with an oath cried, " They," meaning
his brothers, " do not half fine them here.   You should
come over to my place and see my books."   In less than
a week this man was seized with a fit and fell dead ; and
the poor people, in their simplicity, said it was a judgment
sent from Heaven.   The owners of the firm evidently did
not think so, for they continued to fine the people as
usual.

On the following Whit Monday more after-dinner
speeches were being made when the head of the bleaching
department boasted that he had six hundred pounds
which he had mulcted from his workers in fines.   He
forthwith built a new house and furnished it in grand
style, and sent me to Taunton to buy him a grand piano.
The day for opening the new house was appointed,
friends were invited and every preparation made for a
grand feast.   The day arrived, and the young boaster was
riding up and down on his high-mettled white horse inspect-
ing the arrangements when the horse, treading upon a half-
rotten turnip, plunged and fell, and rolled completely over
its rider, who was taken up and carried into the new
house where he had never lived, a corpse.   The simple-
minded people also called this a judgment from above.
Very soon after this a large fire occurred at this factory
which threw a great number of men, women, boys and
girls out of work.   As for myself I soon engaged myself
to a solicitor, and my duties varied between serving writs
and gardening.   Every time I delivered a writ I had to
be put on my oath the following morning that I had so
delivered, and then I used to receive three shillings and

sixpence for my trouble. Sometimes I had to make three or four journeys before I could accomplish the delivery, as so many well-known gentlemen would cleverly keep out of my way. On one occasion I had to serve a writ on a gentleman who always managed to be away when I called ; but my master told me that if I had reasons for believing that he was at home I should walk in. The next evening when I called and the servant told me her master had just gone out, I pushed past her, and got into the dining-room, where the gentleman was just cutting away at a piece of roast beef. I handed the document to him and, bowing politely, retired as hastily as possible. As I turned to the door I heard something whirr past my shoulder and strike the wall beyond. It was the carving knife.

Another difficult case was that of a master carpenter who lived several miles from my master's office. It was summer time, and he went away to his works as soon as it was daylight and was not at home until nine in the evening. I could never find this gentleman at home although I called several evenings in succession, and at last I grew tired of walking so many miles several times a week. As there was a porch at the door I decided to sit down there and wait until he did come home. I sat and dozed until about 3 a.m., when I heard some one moving inside the house. Presently the door opened, and out came the gentleman I wanted. He was thunder-struck to see me there. I delivered the writ into his hands, and we both said " Thank you," and went our ways.

After I had been in this situation two years it was a very disagreeable blow to me to hear that my master was retiring from business and would have no further need of my services. He and I had got on together very comfortably for two years, and he was a kind employer, so I was really very sorry.

I was now twenty-five years of age, and, like other young men, I thought it time to begin to see about " committing matrimony." I had become acquainted with Miss Anne Warner, who lived at Henley-on-Thames, and she decided that she would marry me, if I got a permanent situation either on the railway or in the post office. Accordingly, I applied for the railway, and was appointed as porter at Bristol in May, 1850. The Bristol and Exeter railway at that time was in its infancy. Amongst the articles served out to me was a wooden staff or truncheon, to be used, if necessary, for clearing the station.

In October of the same year I was married to Miss Warner at St. Mary Redcliffe's Church, Bristol. I knew that I was not to remain long at Bristol, so my wife and I took rooms for a while. I was very soon sent as signalman to Stoke Canon, a village in Devonshire, where at that time there was no station, but only a crossing. I had to leave my wife in Bristol until I could obtain a suitable home, which I did as soon as possible, as keeping two homes going was a very expensive thing to do, and my wages at that time were only 16s. a week. My only accommodation when on duty, which was twelve hours a day, or an alternate week's night duty, was a sort of sentry box—a wretched affair, especially in cold weather, and for night duty.

While at Stoke Canon I made a model of the village church. The only tools I had for this work were a pocket knife and a hammer. Some time after I added a peal of bells which were set ringing by putting a penny in a slot in the roof of the church.

In 1853 our eldest daughter was born, and in the following year I was removed to Martock in Somersetshire, where a son was born. In 1857 I was removed to Durston, and in May another daughter was born. My wages were then £1 a week, a very small income on which to keep ourselves and three children and pay rent, but my wife kept a school and had several neighbours' children to teach. We were very happy together, and I was glad that I was born!

GOD IN HIS WORKS

Dost thou love nature ? Dost thou love
Amid her wonders oft to rove,
Marking earth, sea, the heavens above,
    With curious eye ?

Read, then, that open book ; see where
The name of God, inscribed there,
Urges thee on till thou declare,
    " My God, I see ! "

Yet venture not, my soul, to come
Within fair earth's material dome
Without thy God : thou hast no home
    To compass thee.

E

Nature's fair works must e'er be read
As penned by nature's Sovereign Head ;
Else were its loveliest pages dead—
        Without His key.

But by the Polar Star of Grace,
Nature assumes her proper place,
And thou mayst safely lead and trace
        Her harmony.

                                M. M. C.

CHAPTER VII

SCIENTIFIC ACHIEVEMENTS

[By Miss Ellen Langdon]

THE above chapters were written by my father, Roger Langdon, and I, his daughter, Ellen, am continuing the story of his life. So I will begin by saying that the school kept by my mother was conducted in the same manner as were the church schools at that time. Everything was very orderly and we just had to mind our p's and q's.

Our parish church and school were five miles away, so it was only possible for us to go there occasionally. We usually made the journey sitting on a trolley which father pushed part of the way, and then we would clamber up the railway bank and walk on to church. But there was another church nearer, to which as soon as we were big enough father would take us on alternate Sunday mornings, when he was off duty.

We had a large front room which was used as a schoolroom, and here we, with our neighbours' children, were taught reading, writing, arithmetic, spelling, geography, sewing, and the Catechism. When father was on night duty (the duty in those days was twelve hours at a time) he would be at home in the morning, and sometimes he would take us for Scripture. In this his teaching was unorthodox and advanced, and he always gave us plenty to think about. When later on we went to a school at Taunton we found ourselves in most subjects in advance of children who had attended schools in the town.

During the eight years my parents spent at Durston three more sons were born to them ; so there were now six of us, and I have often wondered since how mother managed to keep the school going with such a large family of her own. My grandmother, my mother's mother, was a frequent visitor, and would also be at work all day long. My father's father too would often come, and he used to make us stand by the harmonium and sing. I would like to say that my grandfather was one of Father Matthew's earliest converts to teetotalism, and he tried his best to get others to believe in the same thing.

At this time father made a harmonium, which proved such a success that he was able to sell it, and with the proceeds he bought material to make a second. He also made a magic-lantern, and made slides of the stars and sun and moon and comets, and at Christmas time he would invite the neighbours to see the lantern, and he would give them all the information he could on the subject. He also told us about the electricity which was some day to light up our houses and town and drive our railways and carriages. He told us about photography, and how we should live to see "living pictures." Then for us he made all sorts of mechanical toys—walking dolls, wooden horses and boats. The forms which were used in the schoolroom he also made, as well as tables, chairs, our boots, and his own, a little carriage for the smaller children, and later on a perambulator. He was able to make anything he wished, but of course it all meant labour, and never a moment's idleness. And he had to put up with a great deal of enmity on account of his being so set against the drink traffic, and never going to the *Railway Hotel* to spend his earnings.

While we were at Durston we had the pleasure of seeing Garibaldi. His train was crossed on to another line. We children were playing as usual on the bank above the station, and when Garibaldi's carriage stopped right in front of us of course we all screamed with delight, and our noise brought everybody out, and the men got so excited they crawled all over the top of the carriage and shouted for all they were worth.

In 1865 my father was removed to Taunton. His duty was at Norton Fitzwarren crossing, where there was no station, so he had only a small office to sit in, which, as he still had the night duty, he began to find very trying. House rent at Taunton was heavy, so mother applied for a schoolmistress's situation, and she kept this post for nearly two years. Meanwhile another son was born. But we were not very happy at Taunton, for father was often ill during the time we were there.

The school in which mother taught was next to our house, and was also used on Sundays as a Mission Chapel. When we first went there they used to chant the *Magnificat* and *Nunc Dimittis*, and a hymn, to the accompaniment of an accordion.   Father did not like the sound, especially when he had to sleep on Sunday before going on night duty ; so he offered to lend his new harmonium if they would take care of it and could find some one to play it.   The curate, the Rev. J. Jackson, was much pleased with the idea, and he persuaded a crippled lady, Miss Emma Mockeridge, to undertake the duties of organist.   She used to be wheeled down on Sundays to play, and twice in the week besides she would gather the children round her in the schoolroom and teach them the hymn.   So she came in all weathers from quite a long distance to do this work, and after a time she began to visit the people who attended the service, and told them that since Mr. Langdon would not accept any payment for the use of his harmonium it was their duty to provide themselves with a new one of their own.   Before very long she succeeded in collecting enough money to buy one.   At this time Mr. Jackson held evening classes for men and charged them a halfpenny a week.   Father with some others went to him to learn Greek, and got on very well with it.   He greatly enjoyed these classes, and in later life would say how grateful he was for them.

While we were at Taunton there was an election, and father had a vote.   A great deal of bribery went on ; indeed the Conservative member was afterwards unseated for bribery.   Mr. Henry James (now Lord James) was the Liberal candidate, and he won the election.   My father told us all about it, and said if any one meeting us in the town, or calling at the house, should offer to give us children anything we must on no account take anything.

I do not know who the people were, but they mostly called in the evening, and would offer us groceries and other things. One evening I had been to the station to meet father, when a gentlemanly looking person came up to him, saying, " Mr. Langdon, I believe."

Father raised his hat and said, " Yes, that is my name."

" You have got a vote ? "

" Yes, I have."

" Well, are you going to give it to us ? "

" That is my business, not yours."

" Come now, don't be a fool," said the gentleman, " you have got a little family ; what will you do it for ? " at the same time holding up one finger. " I am not to be bought," father replied. The gentleman then went on holding up fingers till he had got all his ten fingers up, and father at last cried, " No, I tell you no," to which the other replied, " You thundering fool." Father raised his hat with " Thank you, and you are another," and off we went.

My father was deeply interested in astronomy, which he had studied a great deal with the help of books, and he had bought celestial and terrestrial globes. Now he wanted a telescope. Mr. Nicholetts, a dear old gentleman who lived at Petherton, had a telescope ; and he often invited father to his house to look through it, and this gave father great pleasure and increased his ambition to possess one of his own. So for a few shillings he bought some second-hand lenses, and soon succeeded in making a small telescope, with a $1\frac{1}{2}$-inch reflector mounted on a wooden stand and swivel. This small instrument only whetted his desire for something better ; so he sold if for 7s. 6d. and with the money obtained materials for another. After many difficulties and disappointments, which by sheer luck and hard work he surmounted, this second telescope was at last completed. This one had a four-inch reflector, and with its aid the ring and some of the satellites of the planet Saturn could be seen. The crescent form of Venus and some of the nebulæ were also plainly visible. And when father first saw the moon through it he said he was fairly astonished, for up to that time he had no idea how much of the physical features of the moon could be seen.

In 1867 father was appointed station-master at Silverton in Devonshire. It was at the end of 1868 that we left Taunton and took up our abode in our new home ; and thus began what father always described as the happiest time of his life. For one thing he had from this time forward no more night duty, and his health improved considerably in consequence, so that he became stronger than he had been for many years. He greatly rejoiced, too, that there was no drinking bar at this station. Another great advantage was that we were now within reach of Exeter where there was a good school to which the younger children could be sent daily by train. In March 1870 my youngest sister was born ; so now there were eight of us. But the following year my eldest brother was killed by an accident at the station. This was a terrible blow to both my parents, and the trouble turned father's auburn hair as white as snow. At Silverton father made many friends, amongst them Sir Thomas Acland and our good rector, Rev. H. Fox Strangways and his lady.

In 1870 my father became acquainted through the *English Mechanic* with Dr. Blacklock, and this gentleman gave him advice regarding the building of his telescopes ; but it was all done by letter, for they never met, and it was wonderful how Dr. Blacklock found time with all his work to write so many letters as he did. Father also received letters from Mr. Nasmyth, the inventor of the steam-hammer, on the same subject. Mr. Nasmyth took great interest in him and would write two or three sheets at a time, pointing out the difficulties and explaining how they might be overcome, and drawing diagrams of the tools he would need. After father had made the speculum, which he found exceedingly difficult, it had to be silvered on the front surface ; and on this point Dr. Blacklock gave him valuable information which enabled father to do it successfully at the very first attempt.

In 1874 father made a model in plaster of Paris of the visible hemisphere of the moon, showing five hundred principal objects, hollows, craters, and mountains. This model he afterwards presented to the Devon and Exeter Institute. Mr. C. R. Collins of Teignmouth once wrote an article describing the discovery of a new crater on the moon by Dr. Hermann J. Klein of Cologne. Going to his observatory father was able to show on this model of the moon, as the result of his own observations, this very crater.

Father had now made two telescopes, but he hoped to make another and still better one ; so he set to work, and it was in the making of this that he received so much valuable advice from Mr. Nasmyth and Dr. Blacklock. This third telescope was a beautiful instrument. It had a six-inch speculum with a five foot focal length. With this he was enabled to detect certain markings upon the planet Venus. In 1871 he read a paper before the Royal Astronomical Society in London upon this subject. He said afterwards that he never was so nervous in his life as on this occasion, and he wished the earth would open and swallow him up. But his paper was very well received, and commended.[1]   He also made over a thousand drawings and photographs of the moon's surface.

Much as father had accomplished, he was still bitten with the idea that his telescope was not so good as he would like, although it was a splendid one, and had cost him many weary hours' hard work to make. So he sold it for £10, and in the year 1875 set about making his fourth telescope—a very noble instrument, for which he had to build an observatory. He describes it thus : " An 8¼-inch silver-on-glass reflector mounted in a stout zinc tube, which turns in a cast-iron cradle on its own axis. The focal length is seven feet. There is a diagonal place for viewing the stars and a specially prepared glass wedge for observing the sun. The whole is mounted as an equatorial

---

[1] Webb quotes from this paper in his book *Celestial Objects for Common Telescopes*. My father's observations are also mentioned in Clarke's *History of Astronomy During the Nineteenth Century*.

upon a strong cast-iron stand. It had two stout brass
right ascension circles divided to 10 seconds, and declina-
tion circles divided to 5 minutes of arc. The telescope is
furnished with a driving clock which keeps the celestial
object in the field of view. The observatory is a circular
iron building with conical-shaped revolving roof, two swing
flaps of which give the required opening to the sky."

This telescope took a long time to make, and night
after night through many a weary month, when station
duty was done, father would work at it for hours
together in his home-made work-shop. But, as usual, the
want of funds hampered him a good deal, and he found many
difficulties to overcome ; but he worked away with intense
enthusiasm, and with the advice of Dr. Blacklock and Mr.
Nasmyth, he at last completed this Newtonian equatorial
reflecting telescope fitted with a finder with Ramsden
eye-piece. He added to it a trap for taking photographs
the invention of his own brain, and in visiting Greenwich
Observatory some years later he was pleased to find that
the apparatus in use there for the same purpose was
almost identical with his own. With this telescope my
father photographed the transit of Venus and took
also several pictures of the sun and of the moon.

To make his first telescope in 1865 he bought some
second-hand lenses for a few shillings, and by means of a
turning lathe he turned a stick upon which to roll the
tin case, according to the size of the lens.

The second telescope was a much more difficult under-
taking for one whose acquaintance with mechanical pro-
cesses was entirely self-acquired. He was as a man

groping in the darkness. To obtain the special glass necessary for the speculum ; to grind it to the most delicately accurate shape and density ; to polish and silver the speculum ; to make the metal tube to the requisite size and scale ; to mount it with the necessary adjustments and accuracy ; all these were so many enigmas which only his intense enthusiasm and perseverance enabled him to solve.

To grind the speculum of the third telescope a special and very curious tool was necessary, and here Mr. Nasmyth gave father valuable information and sent a drawing of the tool. After making this tool according to Mr. Nasmyth's pattern, father found it applicable to metal specula only, and unsuitable to glass inasmuch as it would not parabolize, or work in the figure of a parabola. This was a great blow. However he eventually surmounted it by using Ross's machine, a description of which he came across ; after considerable inquiry. This grinding completed, he succeeded in polishing his speculum with a disc of pitch squares, an apparatus which gave him much thought and trouble. Then came the silvering, and then the rolling and soldering of the tube, which was accomplished by means of a circular block of wood turned in the lathe to the required size. Here another difficulty presented itself, for when it was done, the wood was found immovably fixed inside the tube, and it became necessary to procure a steel augur to bore it out.

However, this taught father something, for in making the next telescope he used a number of laths fixed together and turned in the lathe. One of them was cut through diagonally so that the two parts when separated formed wedge-shaped sections, which could be readily knocked out when the case had been fixed ; and thus the whole circular bundle could be easily removed. I know that the building of these telescopes was real hard work, and the difficulties and disappointments they involved were numerous, and were only overcome by sheer hard work and indomitable perseverance. The fourth telescope he had reason to be proud of. He was assisted with the adjustment of this as well as in making the pitch plate with which to polish the speculum by Mr. Newton, a gentleman from Taunton. This was the telescope for which father built the observatory, and it has been described as a real triumph of skill.

Many a time after his day's work was done he would take his magic-lantern and give a lantern lecture on astronomy. He also wrote a paper, " A Letter from the Man in the Moon," which was published in the *Exe Valley Magazine*. Another paper, " A Journey with Coggia's Comet," appeared in *Home Words*. Some of his mechanical toys, Stoke Canon Church with its peal of bells, ships rocking on the ocean, and others, have often been shown at church sales of work, and so helped the funds.

CHAPTER VIII

CLOSING YEARS

O N several occasions during the early years at Silverton, my father had trouble with drunken passengers. On one occasion a certain book salesman came to the station and called for a ticket to Exeter, for which he tendered 5d. The parliamentary fare being 7d., my father asked him for the other 2d. The man began to abuse him and got on to the line, and would have been killed by an express, but father jumped down and dragged him back just in time to save both their lives. The man then struck father in the face with his umbrella and swore tremendously. After some trouble father succeeded in placing him outside the station gate and locked him out. The man finally paid 7d. for his ticket, and then threatened to kill father. Of course he was summoned and had to pay heavy fines. Father wrote regarding this case: " If I had caused the death of this man, I should have had to do at least twelve months' hard labour in one of Her Majesty's country mansions, and there would have been two and a half columns in *The Times*, *The Standard*, and *The Daily Telegraph*, expatiating on the carelessness of railway officials ; but having saved his life at great risk of my own, I received as complete and satisfactory a blackguarding as it is possible to conceive."

On another occasion father was not so fortunate in averting disaster. In 1879 a lady came to the station in her carriage to meet some friends who were going to spend Easter with her. It was the day before Good Friday, and the trains were very late. The friends were coming in the down train from London, and she also wanted to see her son-in-law who was passing in the up train. While waiting she constantly crossed the line, first to the down platform, then to the up. A down train was signalled and off she went to the down platform. She was a very genial person and had been chatting pleasantly to every waiting passenger. This train was an express, and as soon as it passed by she saw an up train approaching. She immediately attempted to cross the line, probably thinking it was the stopping train, instead of which it was an express. Father rushed out to warn her, but it was too late, for the engine was upon her, and she was instantly killed. The shock was very great to both of my parents and they could not sleep for weeks.

Up to 1874 father was single-handed, and used often to be on duty from 6 a.m. till 2 o'clock the following morning waiting for a coal train which used to come at any time in those days. Afterwards a signalman was appointed, and then father's duty hours were 6 a.m. till 10 p.m. There are several people still living, both rich and poor, who could record the great courtesy they received at Silverton station. Many a time he would help some poor person along the lanes with her babies or her bundles, or show the way with his lamp to some benighted place perhaps two or three miles distant.

In the autumn of 1875 father was very ill, and when he got better he was ordered away for a change of air, and I had the pleasure of going with him. We visited some relations and went on to London. If we had gone to the dullest place in the world I should have been quite happy so long as father was with me, for on all occasions he was just the same age as his children. But as it was we went to all the interesting places, and I don't know which of us enjoyed things the most.

About this time the Great Western railway company took over what had been known as the Bristol and Exeter railway, and began to lay down narrow gauge-lines. The line near Silverton runs through a valley, picturesque but wet. The station itself is about thirty feet lower than the floor of Exeter Cathedral. For several weeks during this time there had been a great deal of rain and the valley contained more water than usual. Torrents ran down from the hills and flooded the valley of the Culm to a depth of over five feet. A culvert which drains the main part of these hills passes under the railway close to Silverton station, and this became blocked by two hurdles which had been carried down the stream and become fixed in an upright position right across the mouth of the culvert. Consequently, leaves, brushwood, thousands of apples and other rubbish got fixed on one side of the hurdles, completely staying the torrent. The railway was quickly flooded, and at 10.30 p.m., after the station was closed for the night, down came the express known as " Madame Neilson's train," because it conveyed her regularly from London to her Devonshire home. Owing to the work of laying down the narrow-gauge rails there were a great many timbers collected on that part of the line, and these were lifted by the sudden flood, and floated about on the water. Mother went to look at the flood just in time to see the express coming at a speed of sixty or seventy miles an hour, and she wondered if it would get safely through. The thought had scarcely entered her head when she saw the great engine rear itself up, as if it were a real live thing, then as suddenly drop down again, and she knew that it was off the line. The passengers got a shaking, but were otherwise none the worse, not even wetting their feet as they passed over planks laid across from their carriages to the platform. Madame Neilson and Madame Patti were both among the passengers, so here was a lively night for my mother and brothers. There is no railway hotel or other house near, so mother did her best to accommodate all these people, who were dreadfully hungry. They soon ate up all that was in our house, and there they had to wait for a relief train from Exeter. My two young brothers were called out of their beds, to escort two gentlemen to the village of Silverton two miles away. They started off full of excitement, and when they were about a quarter of a mile away the water was nearly up to their necks ; but they all four went on, and my brothers had to try and get some bread for the hungry people. So they arrived in due course wet through and tired out, but they were none the worse the next day. Altogether it

F

was a most exciting night.   Father traced the origin of
the flood and drew up plans, and received the thanks of
the railway company.

Now there are some people, especially those who live in
large towns, who may think that a small country station
is a very dull place to live in, but that is because they have
never tried it.   Apart from such occasional and exciting
events as that just described, the country has interests
and amusements of its own.   When country people are
waiting for a local train, particularly if it is a market
train, and all the passengers, both rich and poor, are more
or less acquainted with each other, every topic is dis-
cussed, and if the station-master has a few minutes to
spare his opinion is sure to be asked.   For example, when
Doctor Temple was appointed Bishop of Exeter, it made
such a stir that it was the talk of every one, and father's
opinion was asked by every passenger.   Father had read
about Dr. Temple, though he had not seen him.   His
reply to their question was always the same : " I rejoice
to know that Dr. Temple is appointed ; such men are
needed in the Church very much indeed.   He will be
the right man in the right place, and he will thoroughly
do his duty, and he will be a hard worker.   Moreover,
he will make the clergy work, and it is a thousand pities
that so many churchmen have not yet realized what a
strong man they will have amongst them."

Then up spake a countryman, " Then, du yu 'old way
un bein' a tay-totler, Mr. Langdon ? "

" Why, of course I do," replied father.   " That is
the essence of the whole matter, and that is just why the

Exeter people are against him ; but I for one am thankful, and I think it a great gain to the Church of England to have at last a bishop who holds such opinions."

" Well now to be sure, Mr. Langdon, I knowed thee was long-headed, and I knowed thee was an ole Liberal, now I knows thee beest an ole tay-totler."

Father was a broad Churchman, and wished for Church reform. He liked to hear good music and used to go regularly on Sunday evenings to Thorverton Church, as the services there were very much to his liking. He used to say that the Church was behind the times, and did not reach the people generally, chiefly because the clergy were all taken from one class, and in many cases they did not understand the poor. They were also educated over the heads of the people. In politics he was a supporter of Mr. Gladstone.

My father had now taken up photography and had made a collapsible dark room that he could carry on his back. He succeeded, after a few failures, in taking some very good pictures of the moon in 1880 ; and in December 1882 he took a good one of the transit of Venus. He also made an instantaneous shutter to his camera, by the help of which he was able to get some very good pictures, notably one of the old broad-gauge train known as the " Flying Dutchman." This was before dry plates were invented. The next thing he made was an excellent camera. A gentleman named Mr. Wellington gave him information both in regard to this and to photography in general, for which father was very grateful.

In the winter of 1881 there was very deep snow, and our house was snowed up to a height of several feet, so that before he could open the station father had to dig his way out. No trains ran that day, and only one up and one down on the two succeeding days. After that the line was pretty clear again.

In 1888 father and mother received another fearful blow by the death of my youngest brother, after a very short illness, at his lodgings in Exeter. After this time father never appeared to be very well, and before long his health entirely gave way.

During these years several visitors came to father's observatory, among them Mr. Clifton Lambert, son of the General Manager of the Great Western Railway. This gentleman wrote a sketch of father's life, which was published in three different papers, *Wit and Wisdom*, June 1889, the *Great Western Railway Magazine*, September 1889, and in the summer number *Western Weekly News*, 1894, just one month before he died. Visitors would call at all hours, in the day to see the spots on the sun, and in the evening and at night to see the moon and stars.

For many a year father had been a wonder to the simple country folk. They could not understand a man devoting his spare time to the study of the heavens, for the mere love of science. They had an idea that he could "rule the planets." Father used to say that he was astonished at the amount of superstition prevailing in the minds of all sorts of people, not only the uneducated. Even well-educated people would ask him if he

could " rule their planets." He would say that he was
ashamed to hear them ask such a question. People
would come from long distances in the dead of night
to have a look through the telescope. When asked how
he had achieved so much, and brought up a large family
in respectability, his answer was : " It is through the
woman that the Almighty gave me ; she has done the
most."

It was in March 1894 that I went home to find father
suffering from a complaint from which he could not hope
to recover. He got gradually worse, and in July I was
sent for again to come at once if I wished to see him alive.
Mother had done all she could do for him, and I was very
much shocked to find him suffering dreadfully. But he
was very cheerful, only longing for his sufferings to end.
It was a painful time, but as we all gathered around his
bed, he often made us laugh by his jokes. Then there
were quieter moments when he prayed the Almighty
to take him. On the evening before he died, he repeated
part of Psalm li. to me and some hymns, and that was the
last. He passed peacefully away in the morning.

Through the kindness of Sir Thomas Acland, he was
buried in the private burial ground of the Acland family,
near his two sons.

So live that thy summons comes to join
The innumerable caravan which moves
To that mysterious realm where each shall take
His chamber in the silent halls of death.
Thou go not, like the quarry slave at night,

Scourged to his dungeon, but sustained and soothed.
By an unfaltering trust, approach thy grave
Like one who wraps the drapery of his couch
About him, and lies down to pleasant dreams

*   *   *   *   *

Speak of me as I am, nothing extenuate,
Nor set down aught in malice.

APPENDICES

APPENDIX I

A List of Parish Clerks of the Parish of Chisleborough, Copied from the Register

Roger Langdon, Doctor of Music, and Clerk of the Parish of Chisleborough, in the county of Somerset, from the year 1769 to 1791. } 1769. 22 years.

His son, James Langdon, held the Clerkship from 1791 until 1822. } 1791. 31 years.

Edward Langdon, grandson of Roger, held the office from 1822 until 1871, also Organist. (Father of the subject of memoir.) } 1822. 49 years.

Edward Langdon, great grandson of Roger, was also Parish Clerk and Organist from 1871 until 1885. } 1871. 14 years.

Peter Langdon next held this same position and died recently. Great-great-grandson of Roger Langdon. } 1885.

APPENDIX II

OBSERVATIONS OF THE PLANET VENUS, WITH A 6-INCH SILVERED
GLASS REFLECTOR. BY R. LANGDON

Communicated by J. Norman Lockyer, F.R.S.

[From the "Monthly Notices" of the Royal Astronomical
Society. Vols. 31–32, June, 1872]

On May 1, 1871, I had a good view of the planet Venus, but
I could not at first see her to my satisfaction as her light was
so bright. She had more the appearance of a miniature sun
than a star; but I put a diaphragm of blackened card in the
eye-piece, and made a small hole through its centre with a
piece of hot wire. I found this arrangement keep out to a
great extent the glaring rays. I also sometimes used a slip
of slightly tinted glass in front of the eye lens; this enabled
me to bring the planet entirely under subjection. Her shape
was that of the moon when a little more than half full. I
distinctly saw a dull, cloudy-looking mark along her bright
limb, curving round parallel to it, and extending nearly across
the disc, each end terminating in a point; joining this at the
eastern extremity was another and darker mark of a club shape,
its small end joining the point of the mark previously described. I watched these marks for half an hour. I saw some
marks again the next evening, but before I could examine
them the planet was hidden behind some clouds. On May 6,
at 7.45 p.m., there was a cloud-like mark extending straight

across the disc, and a club-shaped mark nearly in the centre, with its small end nearly touching the straight cloud. On the western limb another dark mark had made its appearance ; it was not quite so large as the other, and it was not club-shaped ; but its sides were parallel to each other till they approached the straight cloud, when they appeared to divide, each side curving round away from the other. I took much interest in watching these spots, as I had read that it was very doubtful whether any marks had ever been seen on this planet. I called several men to look at them, and they were able to describe them, although they had no previous knowledge or idea of what they were likely to see. One man was very confident it was the moon he was looking at, but when I pointed out to him that the moon was not in the neighbourhood, he said he thought it was the moon, because he could plainly see the dark patches on its surface.

On May 13, at 7.30 p.m., there was a dark mark of a pear shape, extending from near the western edge to two-thirds the distance across the bright disc. This mark was not so dark as those seen on the 1st and 6th, but it was much larger.

On July 28, at 8 p.m., there were visible five dusky marks along the planet's terminator, and one nearly in the centre of the crescent, but they were not so well defined as those before described ; but what seemed to me more remarkable was that the southern horn was rounded off considerably, whilst the northern horn was quite sharp, and ran out to a very fine thread-like point.

On October 13, at 5.45 a.m., I saw Venus as a beautiful little crescent. She was well defined, and both horns were as sharp as the finest pointed needles. I think I detected a dusky cloud-like mark about half way from the centre to the northern horn ; but I am not quite sure about this as I had to leave my telescope before I could complete my sketch.

On October 25, at 8.10 a.m., I was gratified with a sight which I had waited for and longed to see for many years ;

that was to have a good view of Venus by daylight. I now had the longed-for opportunity, and it turned out as I expected. The superior light of the sun overcame that of the planet to such an extent that I was able to see her better than I had ever seen her before. I could now plainly perceive the jagged nature of the terminator, the unevenness of which could not be mistaken; but what was very remarkable, the northern horn was bent in towards the centre of the planet; it appeared as if a notch had been cut in the inside, and a slice cut off from the outside.

I have no idea what was the cause of this appearance; I had never seen it so before, neither do I recollect ever having read of such a phenomenon. I did not perceive any markings on this occasion, but there was a kind of haziness along the whole length of the terminator; but I considered this at the time to have belonged to the terminator rather than to any markings on the disc. The terminator on this occasion was inky black.

On November 9 I saw Venus every half hour during the day up to one o'clock. I made a sketch at 12.20 p.m. I could now distinctly see the jagged terminator; the nature of which was so much like that of the moon as it was possible to conceive; except that if we compare the moon's terminator to a piece of network, that of Venus would be represented by a piece of fine lace. I could also see some thin, cloudy marks on her disc. The southern horn was very sharp; the northern one was a trifle rounded.

I saw Venus on February 5, 1870 (a few days before her inferior conjunction with the sun), and the bright part was an exceedingly beautiful fine crescent; but I and several other people could see the whole body of the planet in the same manner as we see the dark limb of the moon when *Earth-shine* is falling upon it; but I did not make any sketch at the time.

I have observed Venus a great many times besides those mentioned above, having made it my special work to do so, and have on several occasions strongly suspected markings to be visible ; but I have not mentioned them, and have only described those times upon which I have no doubt of what I had seen.

SILVERTON STATION, NEAR CULLOMPTON, DEVON.

(Sketches illustrating this and the following paper can be seen at the Royal Astronomical Society's rooms.)

APPENDIX III

Observations of the Planet Venus, with a 6-inch Silvered Glass Reflector. By R. Langdon

[From the " Monthly Notices " of the Royal Astronomical Society, June 1873. Vol. 33, Page 500.]

On January 2, 1873, there was a cloudy mark, of a semi-circular shape, extending nearly across the disc, and a dark spot in the centre ; the illuminated disc itself was singularly egg-shaped. Bad weather prevented me from constantly observing this planet, as I should like to have done, but on April 17, at 8 p.m., I was viewing the planet with one of Mr. Browning's excellent achromatic eye-pieces, when I saw two exceedingly bright spots on the crescent—one close to the terminator towards the eastern horn, and the other in the centre of the crescent. These spots appeared like two drops of dew ; they were glistening in such a manner as to cause the surrounding parts of the bright crescent to appear dull by contrast. Cloudy weather prevented me seeing the planet again until the 19th, when the spots had disappeared, but the planet on this occasion was seen through the Aurora, and the irregular and uneven appearance of the terminator was most beautifully depicted. The whole body of the planet also was distinctly visible.

APPENDIX IV

A Letter from the Man in the Moon

(Left for the Editor at the Railway Station)

[Reprinted from the *Exe Valley Magazine*]

Dear Cousin,—

Knowing how exceedingly anxious you must be to find out all you can respecting this little planet on which I live, I take this opportunity to send you a few lines to give you some little account of it.

The moon, in many particulars, is like the earth on which you dwell; and perhaps there is no better way to give you a little more information about this planet than by instituting a comparison between it and the earth.

I must presume you are aware that the earth is a globe, nearly round, like an orange; its circumference is about 24,000 miles, and its diameter 8,000. The moon in this respect is like the earth, being also a globe, but it is only 2,160 miles in diameter, and about 7,000 miles in circumference. It would therefore take forty-nine moons to compose a globe the size of the earth. If you will take two threads and suspend an orange and a small cherry at six feet apart, you will then have a fair representation of the relative size and distance from each other of the earth and the moon. But the earth and the moon are not suspended by any visible or tangible object, but were launched forth in the beginning, and are kept in their places by the balance of attraction, constantly revolving, and travelling onward by the direction of Him who also created the insignificant worm, and whose tender care is over all His works.

You know, dear cousin, that the surface of the earth is diversified with large continents, which are dotted with chains of mountains and high hills, some of which are in a state of volcanic eruption. You have also great oceans of water lying in the hollows of your world. In the moon, too, we have mountains and hills, some of them very high and steep, thrown up ages ago by volcanic agency, though at present there is not a trace of existing fire or volcanic action, and you may safely consider the whole mass of the moon to be a huge, exhausted, burnt-out cinder. Your mountains and hills are denuded—worn down—their sharp points and angles are worn away by frost, rain, and snow, and other atmospheric influences which have been constantly acting upon them for ages ; but here in the moon we have no such thing as an atmosphere : we therefore have neither clouds nor rain, nor frost nor snow ; and in the words of the poet—

Here are no storms, no noise,
But silence and eternal sleep.

All here is as quiet and silent as the grave. Sometimes, from the great heat of the sun, great masses of rock will split and crack, and come tumbling down from the sides of the cliffs ; yet if you were close on the spot you would not hear the slightest sound, because there is no atmosphere by which sound can be propagated and conveyed. Your fields are clad with verdure, and your pastures with flocks, so that as one of your inspired poets has sung—

The valley stands so thick with corn
That they do laugh and sing,

but neither verdure nor corn can exist upon the moon, as no plant-life can grow in a vacuum where there is no moisture.

The crater mountains of the moon are its grand peculiarities. We see here that its whole surface has been upturned, convulsed and dislocated with forces of the greatest activity, the results of which remain to this day; so that our rocks are not levelled down by the fury of tempests, nor smoothed by the constant flow of water, as your earthly mountains are, but stand up in all their primitive sharpness. These volcanic craters are of all sizes, from fifty yards to as many miles across, and in the centre of some of them there stand up lofty hills. Now if you could take up your position upon the highest peak of one of these central hills at the time the sun was rising, you would see the tops of the distant mountains forming a circle round you all illuminated by the sun's light; but as there is no atmosphere, there is no twilight, and consequently the great valley immediately beneath your feet would be in the very blackest of darkness.

I know that you and others have often wondered what those dark grey patches are which you can see upon the moon, even with the unassisted eye. Some people call them " The man in the moon, and his bundle of sticks," and the story goes that I went stealing sticks on a Sunday, and for my wickedness was banished (sticks and all) into the moon ! Now I most strongly protest against this cruel libel; I never stole any sticks, even on a week-day, much less on a Sunday, and I must say the people must have dreadful weak eyesight, and a dreadfully strong imagination, to see anything in these dark patches that can possibly be stretched into the shape of a man with a bundle of sticks at his back. So I hope you will kindly contradict this calumnious story whenever you can ; indeed, in writing, it was partly my object to ask you to do so.

One of the smallest dark markings that you can see on the moon with the naked eye is known to selenographers by the name of Mare Crisium, or the Crisian Sea ; its width across from north to south is 280 miles, and its length is 354 miles from east to west, and it contains about 78,000 square miles, more than half as much again as the area of England and Wales—rather a large size for a bundle of sticks, I opine. There are several other dark or grey patches on the moon, some smaller and some larger than the Mare Crisium, but they are all the beds or bottoms of what were once oceans, seas, and lakes, the waters of which have been dried up or evaporated many years ago. Some think they have all gone over to that side of the moon which never turns round towards you, but I can tell you that is not the case ; for if any water did exist on the moon's surface, the attraction of the earth would certainly draw it round to that side nearest to you, and so you would be able to see some signs of it, as well as clouds and vapours which would rise from it during the time of full moon.

There are many other objects of interest, which I could mention to you, but I must draw my letter to a close ; I will therefore only just give you the names of a few of those dark hollows which you can see with the unaided eye when the full moon is shining brightly.

There is the " Sea of Tranquillity " : its width from north to south is 432 miles, and from east to west 425 miles. There are also the " Sea of Serenity," the " Sea of Fogs," the " Frozen Sea," the " Sea of Vapours " and the " Gulf of Rainbows." This last named will appear to you of a greenish tint, and it is surrounded on nearly all sides with very lofty and steep mountains, some of them more than 15,000 feet high. Then there are the " Ocean of Storms," the " Gulf of Dew," and the " Sea of Humours." This last will also appear of a green tint ; it is very level, and is 280 miles across.

Next come the " Sea of Nectar " and the " Sea of Fertility." All these were named " seas," because the ancient astronomers thought they contained water, and that they really were seas ; but you are aware now that they contain no trace of water, so I need not inform you of that fact. And now dear cousin, I sincerely hope that what I have written will interest you, and if it does, and you will kindly let me know, I will write you another letter at some future time ; but for the present I will say—Farewell !

Your faithful servant and attached cousin,

" The Man in the Moon."

R. LANGDON.

APPENDIX V

A JOURNEY WITH COGGIA'S COMET

[Reprinted from *Home Words*]

This comet, which last year excited so much interest, is supposed by some to be the same which appeared in the year 1737. If so, it is beyond the power of the human intellect to calculate the number of miles (millions upon millions) which it has travelled since that date ; we may, however, in imagination, travel with it on one of its journeys.

Starting off then, as soon as it has made its perihelion passage, we are carried in the course of about six months to such a distance that this comparatively insignificant world (of which nevertheless we are all anxious to get a good slice) disappears entirely from our view, and the larger planets of this system are reduced to mere specks of light. The sun itself, which here scorches us at noonday, only appears there as a very minute star, just a small yellow speck. But meantime other suns, some of them of far greater magnitude and superior brilliancy than the sun we have left behind, gradually come into sight, and some of the " nebulæ," which appear to us here as so many bits of faint hazy light, some of them no larger than a crown piece, now appear to our unassisted vision in all the glorious majesty of suns and worlds and systems of worlds, all revolving round each other in the most regular and systematic order ; for, as Milton says in *Paradise Lost*, " Order is Heaven's first law."

After our steed had carried us for the space of about seventy years in a direct onward course through systems of worlds by us from this world unseen, we should begin to return homewards, but by a different route from that by which we went out ; and we should consequently have a constantly varying scene presented to our view. How awfully grand, for instance, would be the change, as we gradually lost sight of our yellow sun, to find ourselves arriving in sight and under the influence of a sun of a rich crimson red colour, and again after a few years to find ourselves in the presence of a green or a blue sun ! Yet it is more than probable that such would be the case, for the sky is spangled with suns of all colours.

In the course of about 137 years from the time when we set out we should be returned sufficiently near to this world to enable its inhabitants to catch sight of our steed's tail. And then, after all this long journey among the stars of 137 years, we should have seen but a mere atom, just one grain of the works of Him who knows the number of the hairs of our heads, and without whom a sparrow doth not fall to the ground !

Suppose we now inquire, What is the comet's probable business in coming amongst us once in 137 years ? Are its duties those of a messenger or a scavenger, or both ? It is well known that the sun is continually giving off light and heat, and consequently it must of necessity be gradually exhausting itself. It has been computed that were the mass of the sun composed of Newcastle coal, with exhaustion going on at the present rate, the whole mass would be burnt out in 25,000 years. If the sun, then, is gradually being exhausted by giving off light and heat to his family of planets, and if the planets cannot give any portion of it back to him, seeing that they are entirely dependent upon the sun for their own physical and material existence how is the sun's strength to be kept up so as to be equal to the demands made upon him ?

It is but reasonable to suppose that the comets (of which the firmament is said to be as full as the sea is of fishes) should bring some subtle fluid of which this system is being exhausted, and at the same time collect and carry away to other systems some noxious gas or other essence of which we have a super-fluity, but which might be quite essential to the well-being of some other system ; and that so a sort of healthy circulation in the universe around us might be kept up.

Ought we not, therefore, to look upon the appearance of a comet with some such feelings (only in a wider sense) as we would hail the arrival of a ship from a long voyage to a foreign clime, feeling sure it must come laden with some good store for the benefit of those whose business it is to stay at home performing faithfully their own several duties in their own several spheres ?

R. LANGDON.

APPENDIX VI

The Planet Venus

My dear Nelly-Bly,—

According to promise I send a sketch showing the different positions of the planet Venus with regard to the earth during the past few months.

I am astonished, not to say grieved, at the very great amount of ignorance and superstition which exists respecting the apparition of this planet recently as a " morning star."

If you will refer to the sketch I will try and point out the various positions.

June 13, 1887, Venus began to appear low down towards the western horizon as an " evening star," but as the evenings were then light I suppose it did not attract public attention. Daily, however, the planet for a time was seen—after sunset—higher and higher in the western sky, until August 16, when it arrived in such a position with respect to the earth that it sent towards us the greatest amount of reflected light that it is possible it can send at any given time. The planet travelling through space in her orbit at the rate of sixty-nine thousand miles an hour overtook the earth (which is travelling in the same direction at fifty-eight thousand miles an hour) on September 21, when she was exactly in a straight line between us and the sun—called astronomically, "inferior conjunction." The moment she passes this point she becomes a morning star. She still moves on and leaves the earth behind, and when she arrives at the position shown on October 28 she is at her greatest brilliancy as a morning star. From this time the planet's distance from us is rapidly increasing, and consequently her apparent size and brilliancy are as rapidly decreasing, and she is soon altogether lost in the rays of the sun and can only be seen by the aid of a telescope.

Venus makes a complete revolution round the sun in 224 days and 17 hours, but as the earth moves in the same direction but at a slower rate the planet overtakes the earth in about nineteen months, when we have her again as an evening and morning star respectively as before, and so on continually.

And this is the Star of Bethlehem which has caused such a stir within the past two months. All sorts of ridiculous speculation and superstitious nonsense have been said concerning it. Verily in this our day of rapid advancement we are almost, if not quite, as ignorant of astronomical matters as were the " wise men " of the East nearly two thousand years ago, or the natives of Zululand of the present day.

I hope I have made this plain to you ; or if there is anything you do not understand, just ask the question and I will endeavour to supply the information.

Yours affectionately,

R. LANGDON.

PS.—It is rather singular that Venus rotates upon her axis in such time as to produce a Leap Year once in four years as with us.—R. L.

APPENDIX VII

SILVERTON,
Jan. 31, 1888.

DEAR BLY,—

I thought I would send a sketch of the eclipse Annie and
I stayed up to see. We had a very beautiful, clear night,
not a cloud to be seen. The moon entered its eastern edge
into the Penumbra at 8.29 and into the dark shadow at 9.30 ;
and at 11 p.m. the moon was in the centre of the shadow or
totally eclipsed, but we could still see it appearing like a
large orange and we could see all the principal craters and
mountains through the shadow, and I was very interested
to watch the stars disappear one after another behind the
edge of the moon. Of course I have not shown the sun in
the sketch because there is not room, but you must imagine
the sun to be about the size of a round table a good distance
away to the left of the earth.

Now I will try and explain to you what the Penumbra
is and how it is produced. It is produced simply because
the sun is so much larger than the earth, and you can make
a little experiment and show Freddy and William how it is
done. Place two lighted candles or lamps about the width
of this paper apart which will represent the sun, then a little
way off on a white cloth on the table place a teacup upside
down which will do very well to represent the earth, then you
will have the black shadow of the Penumbra or half shadow
on each side of it. A sixpence will represent the moon,

which you can slide gradually across the shadows, and you will have the eclipse to a " t."

It is necessary to have two candles or lamps because the sun is so much greater than the earth and the two candles' light combined represents the sun's light. A small orange perhaps would represent the earth as well or better than a teacup, because the shadow of a cup would not run out to a sharp point, but that of an orange would. That would not matter much either way, only Freddy would most likely be asking the question and you would not be able to answer him.

Your affectionate

D<small>AD</small>.